Fundamentals of Acoustic Signal Processing

Fundamentals of Acoustic Signal Processing

Mikio Tohyama

Institute of Computer Acoustics and Hearing
Kogakuin University
#1-24-2 Nishi-shinjuku
Shinjuku-ku
Tokyo 166
Japan

Tsunehiko Koike

NTT Advanced Technology Corporation
Tokyo 180
Japan

SAN DIEGO   LONDON   BOSTON
NEW YORK   SYDNEY   TOKYO   TORONTO

Academic Press
525 B Street, Suite 1900, San Diego, California 92101-4495, USA
http://www.apnet.com

Academic Press Limited
24-28 Oval Road, London NW1 7DX, UK
http://www.hbuk.co.uk/ap/

ISBN 0-12-692660-3

A catalogue record for this book is available from the British Library

Transferred to Digital Printing 2005

98 99 00 01 02 03 MP 9 8 7 6 5 4 3 2 1

Contents

Preface

Most of the fundamentals of modern acoustic engineering, such as communication acoustics, are related to digital signal processing. University-level textbooks on acoustic signal processing are needed as audio and visual technologies are important for multimedia and computer network systems. This book presents a framework of the basis for theoretical acoustics and acoustic signal processing for senior students in undergraduate courses of universities. Engineers and postgraduate students will also be able to recognize and verify fundamental issues in signal processing and acoustics by reading this book.

Recent developments in acoustic technologies such as sound field control and speech processing are based on digital signal processing. Unfortunately, even well-educated acoustic engineers frequently have much difficulty in understanding signal processing technologies. This is probably because recent rapid progress and growth in digital signal processing technologies require a new framework for acoustic educational programs. Why acoustic signals can be represented and analyzed by manipulation of discrete sequences such as the discrete Fourier transform is a typical question that puzzles acousticians.

Rapid changes in technology have meant that engineers who have completed their acoustic signal processing courses are not familiar with classical wave physics. This is particularly true with regard to young students. Some students are quite naive to sound waves or speech signals in spite of having taken courses in signal processing for acoustic signals. 'What is the wavemode?' is a fundamental question posed by signal processing engineers. Wave analysis is useful for qualitatively estimating the responses of acoustic systems such as room transfer functions, while signal processing is a powerful tool for numerical evaluation of the systems.

A suitable textbook of the fundamentals of acoustic signal processing must also be a good book for the basis of acoustics. This book provides the fundamentals of acoustic wave theories as well as discrete signal processing. Acoustic systems of interest to us are described on the basis of wave phenomena. The wave analysis, therefore, is helpful in getting a deep insight into acoustic systems through the discrete models which can be constructed using a finite number of observation data records. Fundamental techniques such as Green's functions and Kirchhoff's integral formula are well summarized in this book by constructing and solving the wave equations so that readers might understand the basics.

The authors have tried to make this book self-contained, going from wave equations through to discrete signal analysis without tears or too many prerequisites from the readers. The authors' task has been primarily the selection of the topics. Only the fundamental issues that the authors have been utilizing in their lecture courses, seminars, research and development activities are described. Accordingly, advanced topics such as multirate signal processing, blind deconvolution, wavelet analysis, and adaptive processing are omitted.

All the illustrations and numerical examples have been prepared in order to help readers to gain easy comprehension of the technical issues. Mathematical equations are not numbered but they are rewritten in several places when it is necessary. This makes it easy for readers to recall mathematical expressions. In addition, many deductive steps are described without omission, as lecturers write equations on a white-board in a lecture room, so that mathematical expressions should be comprehensible to readers without difficulty. If the readers are not interested in these procedures, they can skip those steps.

All the chapters can be read almost independently: the readers should be able to understand the content of each chapter in isolation if they do not require to read the whole book. Some topics overlap several chapters, but the expressions may look different. The authors believe that these different styles of presentation of issues of the same kind will make readers' comprehension well-grounded and thorough.

However the authors still had to anticipate that readers might be somewhat familiar with elementary mathematics related to Fourier and Laplace transforms, complex analysis, and linear algebra. Nevertheless, even readers who are unfamiliar with the Laplace and Fourier transforms will be able to understand the z-transform and discrete Fourier transform described through Chapters 2 to 5. In Chapter 9 theorems on matrices are employed without mathematical proofs. Newcomers to university junior courses are advised to study those elementary topics in parallel with this text. This book might be helpful in motivating new students as to why they have to learn mathematics: acoustic signal processing is a big 'customer' of mathematics and students may well be interested in mathematics, signal processing, and acoustics. Engineers who are interested in a variety of application areas of acoustics are recommended to read *The Nature and Technologies of Acoustic Space*, by M. Tohyama, H. Suzuki, and Y. Ando, Academic Press, London, 1995.

Although most of the issues discussed in this book are fundamental and basic, they have been selected and newly organized by the authors, who would greatly appreciate the comments, suggestions, and questions of readers and professionals with the view of making the book up-to-date and easily understandable. The authors are always open to constructive suggestions and questions.

Acknowledgments

This book is based on lecture courses for undergraduate and postgraduate students at Kogakuin University and Waseda University in 1995 and 1996. It is also organized following short technical seminar courses in Waseda Research Institute of Science and Engineering, Pioneer Acoustic Laboratories, and Yamaha Technical Development Center in 1995. The authors are grateful to the students and seminar audiences for their fruitful and exciting discussions. Without those experiences this book would not be available. The authors thank Professor Dr. K. Shirai (Waseda University) for his offer of those opportunities of giving the seminars.

Selection of the technical issues also owes greatly to deep discussions with the engineers who attended the authors' technical seminar programs at NTT Advanced Technology Corporation. The authors also thank NTT Advance Technology Corporation for allowing us to publish our work.

We thank Dr. S. Furui (NTT Human Interface Laboratories) for his encouragement. We also gratefully acknowledge all of our colleagues and students (listed below in alphabetical order) for their dedicated cooperation in calculations, simulations, drawings and manuscript preparation. Many thanks to Mr. J. Eeghem, Ms. M. Kazama, Ms. M. Kobayashi, Mr. T. Ohnishi, Mr. K. Sada, Mr. S. Shoji, Mr. T. Tagaeto, Mr. M. Uchiyama, Mr. S. Ushiyama, Mr. N. Watanabe, and all the editorial staff of Academic Press, London for their very kind encouragement. This was a genuinely pleasant working period for the authors.

1 Introduction

Acoustic signal processing is based on discrete linear system theory and wave physics. Chapters 2–5 and Chapter 9 deal with the linear discrete system theory, and Chapter 8 describes the properties of continuous models for linear acoustic systems. Chapter 6 provides the fundamental basis of acoustic wave physics and Chapter 7 discusses the statistical models for acoustic transfer functions. Readers who are already familiar with the classical acoustic wave theory are able to skip Chapter 6. On the other hand Chapter 6 can be used as a textbook for a short course of acoustic wave theory. More or less all the chapters are independent of each other.

Acoustic signal processing starts from Chapter 2, which describes discrete signal expressions and the sampling theorem. Readers encounter sequences instead of functions and they will be familiar with pulse-trains and expressions of their spectra. The sampling theorem provides the readers a reliable basis for discrete signal representations and an answer to the question why acoustic signals can be represented and analyzed by manipulation of discrete sequences. Brief introductions to Fourier and Laplace transforms are described. Readers who are unfamiliar with those transforms will find a fundamental grounding in Fourier and Laplace transforms which will be required through this book. Residue calculations are used in this chapter; however, those calculations will be explained in more detail in Chapter 3.

Chapter 3 discusses the z-transform which is a key issue of digital signal processing and discrete system theory. The z-transform is introduced as the Laplace transform for a discrete sequence. Readers will be able to recognize that the z-domain is continuous, even if a signal is represented as a discrete sequence in the time domain. The contour integral and the residue theorem in the complex frequency domain are introduced in this chapter. The convolution theorem, an important theorem in signal processing is described.

The z-transform will remind acousticians of the spectrum deformation due to reflections with time delays from hard walls close to which a microphone is placed. Such a signal transmission property in a linear system will be discussed in Chapter 4, where readers will discover why the z-transform is necessary for discrete linear system theory.

Chapter 4 deals with the discrete linear system theory using transfer functions. Readers who already know the basics of linear and continuous system theory will discover how to get representations for discrete system characteristics. Poles and zeros, system causality, magnitude and phase, minimum-phase and all-pass system decomposition, and inverse filtering are introduced. The authors recommend readers who are interested in comparison between discrete and continuous system representations to read Chapter 8 in parallel.

Chapter 5 describes the properties of the discrete Fourier transform in detail. The Fourier transform is a fundamental tool for discrete signal analysis and representations just as the z-transform is inevitable in the discrete system analysis. One of the most important properties of discrete signal representation is the periodicity in both frequency and time domains. Students and engineers are frequently puzzled by this periodicity, which is produced by making a discrete signal through the process of sampling of a continuous signal. Readers will come to understand the difference between circular and linear convolutions by recognizing the periodicity.

Chapter 6 summarizes the classical acoustic wave theory. 'What is the wavemode?' is the motivation of this chapter. Rapid changes in information engineering and science require a new frame of university curriculum. Consequently, it is probably not easy for students to obtain fundamental knowledge of classical acoustics. Readers will find that wave analysis is useful for qualitatively estimating the responses of acoustic systems, while signal processing is a powerful tool for numerical evaluation of the systems. Wave equations, eigenfrequencies and eigenfunctions, Green's functions, and Kirchhoff's integral formula are summarized in this chapter.

Chapter 7 describes the statistics for the models of transfer functions of acoustic systems such as random sound fields in room acoustics. Transfer function modeling is a fundamental issue for system identification and control such as sound field control. Acoustic system parameters are in principle determined following the wave equation. However, a statistical approach is frequently taken as a realistic method in practical situations owing to the great deal of variation in acoustic systems of interest to us. This chapter investigates the statistical nature of acoustic systems. Distribution statistics of poles and zeros, random polynomial residue models for room transfer functions, reverberation statistics, and phase statistics of room transfer functions are discussed.

Chapter 8 covers linear acoustic system theory, particularly for inverse filtering. Topics in this chapter overlap those of Chapter 4; however, expressions in this chapter are mainly for continuous acoustic systems using numerical examples. This chapter might therefore serve as a useful reminder to readers of the classical linear system theories. Cepstrum decomposition, minimum-phase, and all-pass decomposition will be discussed in detail. Applications of inverse filtering such as sound image projection systems and source waveform recovery are introduced.

Acoustic engineers frequently feel that recent articles about signal processing issues are too advanced to be appreciated and that discussions proceed too fast using much unfamiliar notation. This is probably because recent rapid progress and the growth in digital signal processing technologies require a new framework for acoustic education programs. One of the fundamentals for the new wave of acoustics is linear algebra, such as simultaneous equations and matrix algebra. Chapter 9 provides a brief introduction to matrix representations for signal analysis and inverse filtering.

Matrix algebra is a fundamental tool for discrete signal and system modeling. A discrete sequence with finite number of data records can be represented by a vector instead of a function of a continuous variable such as time. Convolution between two vectors can be expressed as a matrix which is called the convolution matrix. Readers will be shown in Chapter 9 that both signal analysis and inverse filtering are equivalent to solving a set of simultaneous equations. One of the ways to understand discrete signal processing is to become familiar with a method for solving the simultaneous equations which can not be solved exactly. The least square error (LSE) method is a powerful tool for signal analysis and inverse filtering.

The pseudo-inverse of a matrix using singular value decomposition is introduced for generalizing the LSE approach in Chapter 9 as well as the generalized linear or orthogonal regression analysis for random data statistics. Most of the theorems and properties related to matrix algebra are used without proofs in this chapter; however, readers will recognize how closely linear algebra is related to acoustic signal processing. Although Chapter 9 should be a clear introduction to modern discrete acoustic signal processing, readers who have interest in advanced topics such as adaptive signal processing or the theoretical fundamentals of matrix manipulations are referred to standard textbooks of linear algebra.

2 Discrete Representation of Signals

2.0 Introduction

In this chapter we present the representations of discrete signals and systems. An arbitrary discrete-time signal is expressed as a sequence of numbers using a weighted and time-shifted sum of unit impulses $\delta(n)$. Linear and time-invariant discrete systems are defined and the convolution operation carried out between the impulse response sequence and input sequence is presented.

We present a brief description of the Fourier series expansion of a periodic time sequence and the Fourier and Laplace transforms of transient time sequences. We also present examples of Fourier transforms of several useful time sequences. Finally, the sampling process for continuous signals and the aliasing problem caused by sampling are described. A continuous signal is shown to be reconstructed from only sample values sampled at discrete-time instances using an ideal low-pass filter under a condition imposed on the sampling time interval.

2.1 Continuous vs. Discrete

An acoustic signal is observed as a continuous-time function $s(t)$ where t is a continuous variable representing time. In Fig. 2.1.1 some examples of continuous acoustic signals such as a pure tone, a room impulse response and speech are shown. The objects of acoustic signal processing include controlling the sound field, identifying acoustic systems and synthesizing the spoken word.

Two ways of processing signals are available: one is by using analog circuitry and the other is by digital operation using computer software or digital logic circuitry. Recent advances in high-performance computer technology have made possible more complex signal processing by computers or digital signal processors than was possible using analog circuits. In addition, digital processing can perform tasks which analog circuit could not.

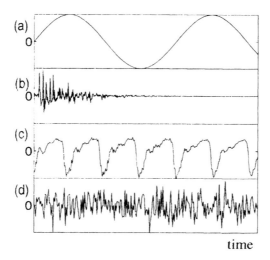

time

Figure 2.1.1 Examples of acoustic signals: (a) pure tone (periodic signal); (b) room impulse response; (c) speech (vowel) and (d) white noise.

In order to process the observed continuous-time signal $s(t)$ on a digital basis, the continuous variable t must be discrete, i.e., the function $s(t)$ has nonzero values only at discrete-time instants. The process is called 'sampling' and results in replacing $s(t)$ by $s_d(nT_s)$ with n integer indicating time index as

$$s_d(nT_s) \equiv \{s(nT_s)\} = \{s(t)|_{t=nT_s}\} \equiv s(n),$$

where we define $s(n)$ or $s_d(nT_s)$ as a sequence of numbers the amplitude of which is given by $s(nT_s)$.

The sequences of numbers $s(n)$ cannot be given to a digital process because the magnitude of $s(n)$ is still of continuous value, so that $s(n)$ must be digitized (quantized) by a further process called analog-to-digital conversion (by an A/D converter). Thus a continuous-time signal can be entered into a digital processing system after discretization in both time and amplitude. The output of a digital processing system, hardware or software, may sometimes be in digital form and sometimes as continuous-time output signals. In order to get continuous-time output signals the digital output signal must be changed back into a continuous-time signal by the inverse process of analog-to-digital conversion (by a D/A converter).

2.2 Discrete Signals and Systems

2.2.1 Discrete Signals

A discrete signal is a sequence of numbers $s(n)$ where n is an index indicating time. It can be obtained by sampling continuous-time signals in acoustic signal processing. The notation $s(n)$ indicates a signal value at the time n and also a sequence of numbers with time variable n. An arbitrary sequence $s(n)$ is expressed as

$$s(n) = \sum_{k=-\infty}^{\infty} s(k)\delta(n-k),$$

where $\delta(n)$ is the unit impulse function defined by

$$\delta(n) = \begin{cases} 1 & n = 0 \\ 0 & n \neq 0. \end{cases}$$

Examples of discrete signals are shown in Fig. 2.2.1 (a) to (d).

2.2.2   Linear Time-invariant Discrete Systems

A discrete system is a system which transforms an input sequence $x(n)$ into an output sequence $y(n)$. The output sequence is called the system response to the input sequence $x(n)$. A system can be characterized by the unit impulse response sequence $h(n)$, which is the system response to the input unit impulse $\delta(n)$. A class of systems called linear and time-invariant systems are particularly important and we describe their properties.

Consider a discrete system which has a unit impulse response $h(n)$, which we call the impulse response for short, and which transforms the input sequence $x(n)$ to the output sequence $y(n)$ as depicted in Figs. 2.2.2 and 2.2.3. Suppose that the sequences $y_1(n)$ and $y_2(n)$ are the output sequences when $x_1(n)$ and $x_2(n)$, respectively, are the inputs. The system is linear if the output sequence

$$a_1 y_1(n) + a_2 y_2(n)$$

is obtained as the response to the input sequence

$$a_1 x_1(n) + a_2 x_2(n)$$

for any constants a_1 and a_2.

Linearity is a property based on the principle of superposition. If the system is linear and its impulse response is $h(n)$, we can easily relate the input sequence $x(n)$ and output sequence $y(n)$. Since the sample value at $n = k$ is $x(k)$, the output sequence $y_k(n)$ at time n to the kth sample input is

$$y_k(n) = x(k)h(n-k).$$

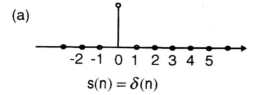

(a)

$$s(n) = \delta(n)$$

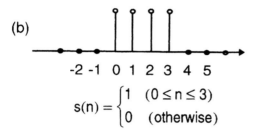

(b)

$$s(n) = \begin{cases} 1 & (0 \le n \le 3) \\ 0 & (\text{otherwise}) \end{cases}$$

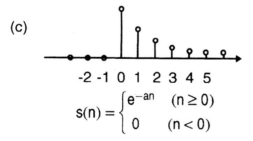

(c)

$$s(n) = \begin{cases} e^{-an} & (n \ge 0) \\ 0 & (n < 0) \end{cases}$$

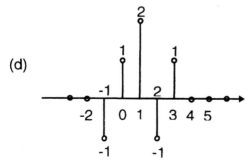

(d)

$$s(n) = -\delta(n+1) + \delta(n) + 2\delta(n-1) - \delta(n-2) + \delta(n-3)$$

Figure 2.2.1 Some examples of sequences: (a) a unit impulse; (b) an impulse sequence; (c) an exponential sequence; (d) an arbitrary sequence.

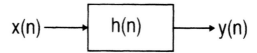

Figure 2.2.2   A system having a (unit) impulse response sequence $h(n)$ where $x(n)$ is an input sequence and $y(n)$ denotes an output sequence.

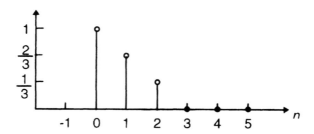

Figure 2.2.3   An example of an impulse response $h(n)$.

Assuming n is a fixed time instant, the output $y(n)$ at time n caused by all the input samples $x(k)$ is the sum of $y_k(n)$:

$$y(n) = \sum_{k=-\infty}^{\infty} y_k(n) = \sum_{k=-\infty}^{\infty} x(k)h(n-k).$$

Here the sequence $h(n-k)$ serves as the weighting constants appearing in the above definition of linearity.

Thus the system output $y(n)$ is expressed in terms of the input sequence $x(n)$ and the impulse response $h(n)$. The above operation to obtain the output $y(n)$ is referred to as a *convolution* and the sequence obtained by a convolution is called a *convolution sum*. We must be sure that the impulse response $h(n)$ does not change depending on time in order to have a unique convolution sum. That is, the sequence $h(n-k)$ is merely the k-sample time-shifted sequence of $h(n)$. The sequence itself is preserved. This property of the system is called 'time-invariance'.

We have another expression of the convolution sum if we reverse the roles of input sequence $x(n)$ and the impulse response $h(n)$. Consider a system with the impulse response $x(n)$. The output sequence $\hat{y}_k(n)$ corresponding to the input sample $h(k)$ at time k equals

$$\hat{y}_k(n) = h(k)x(n-k).$$

Therefore the output sample value $\hat{y}_k(n)$ at time n is the sum of $\hat{y}_k(n)$ caused by $h(k)$ and can be written as

$$\hat{y}(n) = \sum_{k=-\infty}^{\infty} y_k(n) = \sum_{k=-\infty}^{\infty} h(k)x(n-k).$$

Here, changing the time variable as

$$m = n - k$$

leads to the result

$$\hat{y}(n) = \sum_{m=-\infty}^{\infty} h(n-m)x(m) = y(n).$$

This shows that the convolution sum gives us an identical result whichever sequence, $x(n)$ or $h(n)$, is convolved with the other. The notation $*$ is often used to indicate taking the convolution of two sequences $h(n)$ and $x(n)$:

$$y(n) = h(n) * x(n) = x(n) * h(n).$$

Figure 2.2.4 shows an example of the convolution sum. The input sequence is an impulse sequence of four samples all with amplitude of unity (a). The impulse response sequence is shown in Fig. 2.2.4(b). If we consider the nth output sample $y(n)$ (n fixed), $h(n-k)$ works as the weighting coefficients for the input samples $x(k)$; that is, $y(n)$ is obtained as a product of input sequence and $h(n-k)$, sample by sample. Since n is supposed to be fixed and k variable, the sequence $h(n-k)$ becomes a time-reversed sequence of $h(k-n)$ with respect to n. Thus the convolution sum is carried out to get the output sequence $y(n)$ for all n. The convolution sum above is often called a *linear convolution* in order to distinguish it from the circular convolution which is defined for the convolution of periodic sequences discussed in Chapter 5.

2.2.3 Basic Structure of Discrete Systems

A linear and time-invariant discrete system the input–output relation of which is described by a constant coefficient difference equation is built using delay elements, multipliers, and adders. These elementary components are shown in Fig. 2.2.5. Denoting the input sequence as $x(n)$ and the output sequence as $y(n)$, the functions of these components are expressed as

$$y(n) = x(n-k): \quad k\text{-sample time delay},$$

$$y(n) = ax(n): \quad \text{multiplication by } a,$$

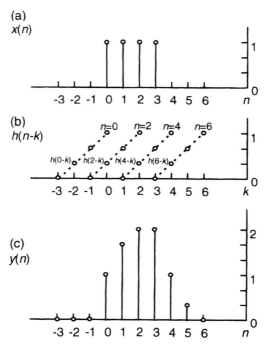

Figure 2.2.4 An example of the convolution sum: (a) an input sequence; (b) weighting coefficients derived from the impulse response $h(n)$ in Fig. 2.2.3 for several fixed values of n where $n = 1$, 3, 5 are omitted for clarity; (c) convolution sum $y(n)$.

and

$$y(n) = \sum_i x_i(n) : \quad \text{addition of input sequences } x_i(n),$$

where i is an index indicating the ith input sequences.

Consider a system represented by the constant coefficient difference equation

$$y(n) = x(n) + \tfrac{1}{2}x(n-1) + \tfrac{1}{4}x(n-2).$$

If we assume that the initial conditions imposed on the system are $x(-1) = 0$, $x(-2) = 0$, and that this system is linear, we have the output sequence $y(n)$ for the input $x(n)$ as

$$y(0) = x(0),$$

$$y(1) = x(1) + \tfrac{1}{2}x(0),$$

$$y(2) = x(2) + \tfrac{1}{2}x(1) + \tfrac{1}{4}x(0),$$

$$\vdots$$

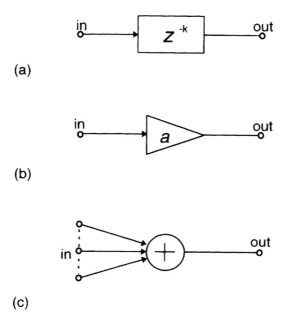

Figure 2.2.5   Fundamental components of discrete systems: (a) a k-sample-time delay element; (b) a multiplier in which input samples are multiplied by a; (c) an adder which has n inputs and one output signal.

The impulse response $h(n)$ of this system is the system response to the unit impulse input $x(n) = \delta(n)$. Assuming the initial conditions above, we get a finite impulse response sequence as

$$h(n) = \delta(n) + \tfrac{1}{2}\delta(n-1) + \tfrac{1}{4}\delta(n-2).$$

A system with a finite-length impulse response sequence is referred to as an FIR (finite impulse response) system. The system configurations of the system using the elementary components mentioned above are shown in Fig. 2.2.6.

Next we consider another system represented by the difference equation

$$y(n) = x(n) + \tfrac{1}{2}y(n-1).$$

The present output sample $y(n)$ is determined by the present input $x(n)$ and the one-sample-past time output sample $y(n-1)$ in this case. We have the output sequence $y(n)$ for the input $x(n)$ as

$$y(0) = x(0) + \tfrac{1}{2}y(-1),$$
$$y(1) = x(1) + \tfrac{1}{2}y(0),$$

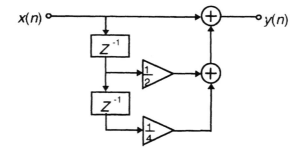

Figure 2.2.6   Configurations of the system represented by the difference equation $y(n) = x(n) + \frac{1}{2}x(n-1) + \frac{1}{4}(n-2)$.

and

$$y(2) = x(2) + \tfrac{1}{2}y(1)$$

$$\vdots \qquad \vdots$$

The impulse response $h(n)$ is the system response to the unit impulse input $x(n) = \delta(n)$. Assuming the initial condition $y(-1) = 0$, we get an infinite sequence as

$$h(0) \equiv y(0) = \delta(0) = 1,$$

$$h(1) \equiv y(1) = \delta(1) + \frac{1}{2}\delta(0) + \frac{1}{2^2}y(-1) = \frac{1}{2},$$

$$h(2) \equiv y(2) = \delta(2) + \frac{1}{2}\delta(1) + \frac{1}{2^2}\delta(0) + \frac{1}{2^3}y(-1) = \frac{1}{2^2},$$

$$\vdots$$

$$h(n) \equiv y(n) = \frac{1}{2^n}.$$

That is we have the sequence

$$h(n) = \begin{cases} \dfrac{1}{2^n} & n \geq 0 \\ 0 & n < 0. \end{cases}$$

A system with an infinitely long impulse response needs an infinite number of delay elements, so it is not realizable. If we use delay elements for the output samples, then the system can be built. In Fig. 2.2.7 the system configuration is shown. Such a system is called an IIR (infinite impulse response) system.

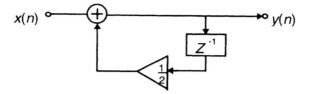

Figure 2.2.7   Configuration of the system represented by the difference equation
$$y(n) = x(n) + \tfrac{1}{2}y(n-1).$$

Consider a general system represented by a difference equation as

$$\sum_{k=0}^{K} a_k\, y(n-k) = \sum_{l=0}^{L} b_l x(n-l)$$

where a_k, b_l are constants and $a_0 = 1$. This difference equation represents an IIR system. The output sequence $y(n)$ of this system is dependent on the past values of itself and is of infinite length. If $a_k = 0$ for all $k \neq 0$, then this system reduces to an FIR system.

Figure 2.2.8 shows a system configuration obtained by the direct implementation of the above difference equation which can be modified to a pair of difference equations as

$$w(n) = \sum_{l=0}^{L} b_l x(n-l)$$

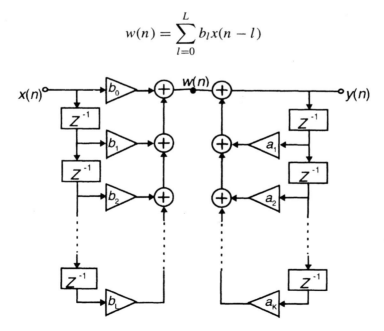

Figure 2.2.8   The direct form-I implementation of a system represented by a high-order constant coefficient difference equation.

and

$$y(n) = w(n) - \sum_{k=1}^{K} a_k y(n - k),$$

where we introduce an intermediate sequence $w(n)$ and

$$\sum_{k=0}^{K} a_k y(n - k) = a_0 y(n) + \sum_{k=1}^{K} a_k y(n - k) = y(n) + \sum_{k=1}^{K} a_k y(n - k).$$

This structure of the system expressed by a constant coefficient linear difference equation is referred to as the direct form-I implementation.

In Fig. 2.2.8 we can see that the structure of the system is composed of two parts which are cascaded. As we showed in Section 2.2.2, we can change the order of convolution of an input sequence and the impulse response of a linear system. Since both parts of the system are linear, we can interchange the connection order of the two parts while keeping the system characteristic unchanged.

Figure 2.2.9 shows another structure of the system obtained by interchanging the order of the right and left parts in Fig. 2.2.8. We find in Fig. 2.2.9 that the contents of the delay elements in the left-half and right-half parts of the system are always identical. So we can combine these delay elements and can get a new system structure as shown in Fig. 2.2.10 with a smaller number of delay elements than that of the system in Fig. 2.2.8. The system configuration as shown in Fig. 2.2.10 is called the direct form-II implementation.

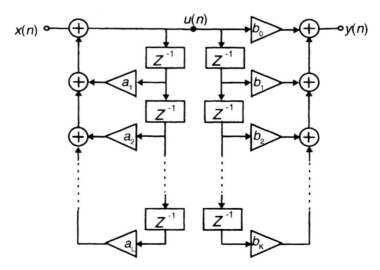

Figure 2.2.9 An equivalent structure to Fig. 2.2.8.

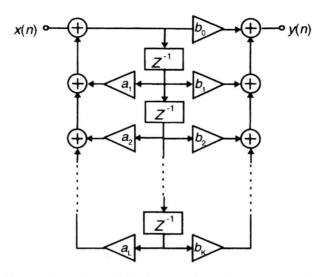

Figure 2.2.10 A direct form-II implementation of a system equivalent to the system shown in Fig. 2.2.9 where $L = K$.

2.3 Fourier Representation of Discrete Signals

The Fourier series expansion and the Fourier transform of discrete signals are useful for the representation of signals in the frequency domain. We present a brief description of the Fourier analysis of signals with several examples in this section.

2.3.1 Fourier Series Expansion

Consider a periodic time function $x(t)$ with a period T. The Fourier coefficients $X(m)$ are defined by

$$X(m) = \frac{1}{T} \int_{-T/2}^{T/2} x(t) e^{-j\omega_0 mt} \, dt,$$

where

$$\omega_0 = \frac{2\pi}{T}.$$

Using the coefficients $X(m)$ we can represent $x(t)$ as

$$x(t) = \sum_{m=-\infty}^{\infty} X(m) e^{j\omega_0 mt} = \sum_{m=-\infty}^{\infty} X(m) e^{j(2\pi/T)mt}.$$

This is called the Fourier series expansion of $x(t)$ and $X(m)$ is called the *frequency spectrum*.

A similar representation is possible for a periodic discrete sequence $x(n)$. Suppose that

$$x(n) \equiv x_d(nT_s) \equiv \{x(t)|_{t=nT_s}\} = \{x(nT_s)\}$$

where T_s is the sampling interval of $x(t)$. We can form a sampled function $x_d(t)$ which represents the sequence using the delta function such as

$$x(n) \equiv x_d(nT_s) \equiv \{x(nT_s)\} \equiv x_d(t) \equiv \sum_{n=-\infty}^{\infty} x(t)\delta(t - nT_s)$$

$$= \sum_{n=-\infty}^{\infty} x(nT_s)\delta(t - nT_s).$$

That is, we extend the sequence into the discontinuous function by padding zeros in the intervals between adjacent samples.

Since the function $x_d(t)$ is periodic with period $T = NT_s$, we can get the Fourier series expansion of $x_d(t)$ as

$$x_d(t) = \sum_{m=-\infty}^{\infty} X_d(m)e^{j\omega_0 mt}$$

and

$$X_d(m) = \frac{1}{T} \int_{-T/2}^{T/2} x_d(t)e^{-j\omega_0 mt}\, dt$$

$$= \frac{1}{T} \int_{-T/2}^{T/2} \sum_{n=0}^{N-1} x(nT_s)\delta(t - nT_s)e^{-j\omega_0 mt}\, dt$$

$$= \frac{1}{T} \sum_{n=0}^{N-1} x(nT_s)e^{-j\omega_0 mnT_s}.$$

If we normalize as $T_s = 1$, then we get

$$X_d(m) = \frac{1}{N} \sum_{n=0}^{N-1} x(n)e^{-j(2\pi/N)mn}$$

since

$$T = NT_s = N$$

and

$$\omega_0 = \frac{2\pi}{T} = \frac{2\pi}{N}.$$

Thus $X_d(m)$ is also a periodic sequence with period $T = NT_s = N$.
In the expansion formula of $x_d(t)$,

$$x_d(t) = \sum_{m=-\infty}^{+\infty} X_d(m)e^{j\omega_0 mt} \equiv x_d(nT_s)$$

$$= \begin{cases} x(nT_s) = \sum_{m=-\infty}^{+\infty} X_d(m)e^{j\omega_0 mnT_s} & \text{for } t = nT_s \\ 0 & \text{otherwise,} \end{cases}$$

where $x_d(t)$ represents the discrete sequence obtained by sampling $x(t)$ at
$t = nT_s$. If we normalize as $T_s = 1$, we get

$$x(n) \equiv x_d(nT_s) \equiv \{x(nT_s)\} = \sum_{m=-\infty}^{\infty} X_d(m)e^{j\omega_0 mnT_s}$$

$$= \sum_{m=-\infty}^{\infty} X_d(m)e^{j\omega_0 mn},$$

where $X_d(m)$ is called the *frequency spectrum* and

$$\omega_0 = \frac{2\pi}{T} = \frac{2\pi}{NT_s} = \frac{2\pi}{N} \qquad (T_s = 1).$$

Since T_s is normalized to 1, notice that ω_0 has no dimension.
The pair of relations

$$x(n) = \sum_{m=-\infty}^{\infty} X_d(m)e^{j(2\pi/N)mn} \qquad \text{for all } n$$

and

$$X_d(m) = \frac{1}{N} \sum_{n=0}^{N-1} x(n)e^{-j(2\pi/N)mn} \qquad \text{for all } m$$

are formal expressions, as the time sequence is written as an infinite series;
however, this pair of expressions is closely related to the discrete Fourier
series (DFS) and the discrete Fourier transform (DFT) pairs which will
be discussed in Section 5.2 of Chapter 5.

2.3.2 Fourier Transform

Let $x(t)$ be a continuous-time signal. The Fourier transform is defined by

$$X(\omega) = \int_{-\infty}^{\infty} x(t)e^{j\omega t}\, dt$$

assuming that the integral exists. Using $X(\omega)$ the signal $x(t)$ is expressed as

$$x(t) = \frac{1}{2\pi} \int_{-\infty}^{\infty} X(\omega)e^{j\omega t}\, dt.$$

A similar relation can be found for a sampled function $x_d(t)$ as

$$x_d(t) \equiv \sum_{n=-\infty}^{\infty} x(t)\delta(t - nT_s),$$

where T_s is the sampling time interval. The discrete-time Fourier transform of $x(n) \equiv x_d(t)$, denoted by $X_d(\omega)$, is defined as

$$X_d(\omega) = \int_{-\infty}^{+\infty} x_d(t)e^{-j\omega t}\, dt = \int_{-\infty}^{+\infty} \sum_{n=-\infty}^{+\infty} x(t)\delta(t - nT_s)e^{-j\omega t}\, dt$$

$$= \sum_{n=-\infty}^{+\infty} \int_{-\infty}^{+\infty} x(t)\delta(t - nT_s)e^{-j\omega t}\, dt = \sum_{n=-\infty}^{+\infty} x(nT_s)e^{-j\omega nT_s},$$

where $x(nT_s) = x(t)|_{t=nT_s}$. If we normalize $T_s = 1$, then we have

$$X_d(\omega) = \sum_{n=-\infty}^{\infty} x(n)e^{-j\omega n},$$

where $X_d(\omega)$ is called the *frequency spectrum density* of the sequence and is a periodic function of the period defined by $-\pi \le \omega < \pi$. This is because we have the relation

$$X_d(\omega + 2l\pi) = \sum_{n=-\infty}^{\infty} x(n)e^{-j(\omega+2l\pi)n} = e^{-j2l\pi n} \sum_{n=-\infty}^{\infty} x(n)e^{-j\omega n}$$

$$= X_d(\omega),$$

where l denotes an integer.

The expression above is interpreted in such a way that $X_d(\omega)$ can be expressed by a Fourier series expansion the coefficients of which are $x(n)$ since $X_d(\omega)$ is a periodic function with period 2π. Using $X_d(\omega)$ we can express the time sequence $x(n)$ as

$$x(n) = \frac{1}{2\pi} \int_{-\pi}^{\pi} X_d(\omega)e^{j\omega n}\, d\omega$$

when T_s is normalized to unity. Here $x(n)$ and $X_d(\omega)$ obtained above are called a discrete-time Fourier transform pair. $X_d(\omega)$ is often written as $X_d(e^{j\omega})$ to indicate that it is a periodic function of ω.

2.3.3 Examples of Fourier Representations of Discrete Signals

We show some examples of the Fourier transforms of discrete signals with illustrations.

(1) Unit impulse (Fig. 2.3.1)

$$x(n) = \delta(n),$$
$$X_d(e^{j\omega}) = 1.$$

(2) Delayed unit impulse (Fig. 2.3.2)

$$x(n) = \delta(n - n_0),$$
$$X_d(e^{j\omega}) = e^{-j\omega n_0}.$$

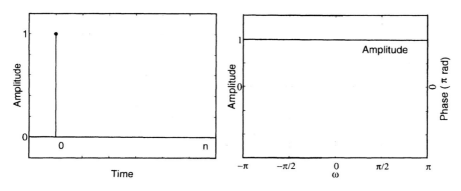

Figure 2.3.1 Unit impulse.

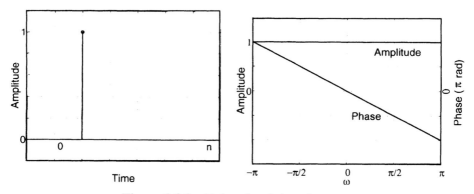

Figure 2.3.2 Delayed unit impulse.

(3) Unit impulse sequence (Fig. 2.3.3)

$$x(n) = \sum_{k=0}^{N-1} \delta(n-k),$$

$$X_d(e^{j\omega}) = e^{-j\omega(N-1)/2}\frac{\sin(\omega N/2)}{\sin(\omega/2)}.$$

(4) Exponential sequence (Fig. 2.3.4)

$$x(n) = \begin{cases} e^{-\delta n} & n \geq 0 \ (\delta > 0) \\ 0 & n < 0, \end{cases}$$

$$X_d(e^{j\omega}) = \frac{1}{1 - e^{-(\delta + j\omega)}}.$$

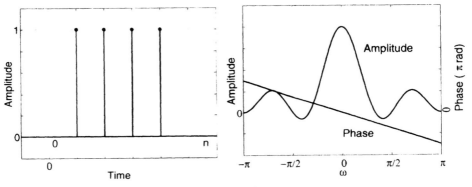

Figure 2.3.3   Unit impulse sequence.

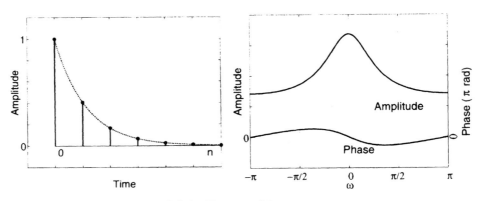

Figure 2.3.4   Exponential sequence.

(5) Exponential weighted sinusoidal sequence (Fig. 2.3.5)

$$x(n) = \begin{cases} e^{-\delta n} \sin \omega_0 n & n \geq 0 \; (\delta > 0) \\ 0 & n < 0, \end{cases}$$

$$X_d(e^{j\omega}) = \frac{re^{-j\omega} \sin \omega_0}{1 + r^{-2}e^{-j2\omega} - 2r^{-1}e^{-j\omega} \cos \omega_0},$$

where $r = e^{-\delta}$.

(6) Sinc function (Fig. 2.3.6)

$$x(n) = \frac{\sin \omega_0 n}{\pi n} \qquad \text{for all } n,$$

$$X_d(e^{j\omega}) = \begin{cases} 1 & |\omega| < \omega_0 \\ 0 & \omega_0 \leq |\omega| \leq \pi. \end{cases}$$

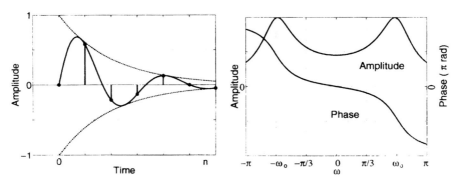

Figure 2.3.5   Exponential weighted sinusoidal sequence.

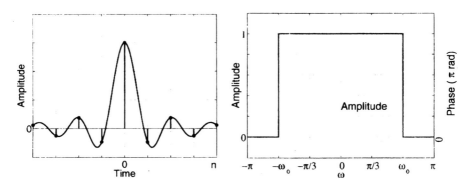

Figure 2.3.6   Sinc function.

(7) Complex exponential sequence (Fig. 2.3.7)

$$x(n) = e^{j\omega_0 n} \qquad \text{for all } n,$$

$$X_d(e^{j\omega}) = \sum_{k=-\infty}^{\infty} 2\pi\delta(\omega - \omega_0 + 2\pi k).$$

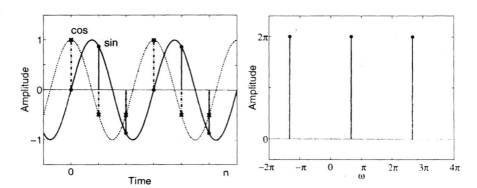

Figure 2.3.7   Complex exponential sequence.

2.3.4   Laplace Transform

The Laplace transform of a time function $x(t)$ is defined by

$$X(js) = \int_{-\infty}^{+\infty} x(t)e^{-jst}\, dt$$

if the integral exists. The variable s is an extended frequency given by

$$s = j\delta + \omega,$$

where ω is the real angular frequency variable. Using $X(js)$, $x(t)$ can be given as the integral

$$x(t) = \frac{1}{2\pi} \int_{j\delta-\infty}^{j\delta+\infty} X(js)e^{jst}\, ds.$$

The integral representation of $x(t)$ is called the *inverse Laplace transform*. The Fourier transform is obtained as

$$X(j\omega) = X(js)|_{s=\omega} = X(js)|_{\delta=0}.$$

Example 2.3.1

The Laplace transform $X(js)$ of the function given by

$$x(t) = \begin{cases} e^{-\alpha t} & t \geq 0 \text{ and } \alpha > 0 \\ 0 & t < 0 \end{cases}$$

is

$$X(js) = \int_0^{+\infty} e^{-\alpha t} e^{-jst} \, dt = \int_0^{+\infty} e^{-(\alpha+js)t} \, dt = \int_0^{+\infty} e^{-(\alpha-\delta+j\omega)t} \, dt$$

$$= \frac{1}{\alpha - \delta + j\omega} = \frac{1}{\alpha + js}$$

if $\alpha - \delta > 0$ where $s = j\delta + \omega$.

The inverse Laplace transform is obtained by contour integral as

$$x(t) = \frac{1}{2\pi} \oint_c \frac{1}{\alpha + js} e^{jst} \, ds = \frac{1}{2\pi j} \oint_c \frac{1}{s - j\alpha} e^{jst} \, ds$$

$$= e^{-\alpha t},$$

where the integral path c is the loop consisting of the real frequency axis and the half circle $Re^{j\theta}$ ($R \to \infty$, and $0 \leq \theta \leq \pi$), and $s = j\alpha$ is only one singularity of $X(js)$ inside the loop.

For a sampled function

$$x_d(t) = \sum_{n=-\infty}^{\infty} x(t)\delta(t - nT_s)$$

we can find the Laplace transform $X_d(js)$ as

$$X_d(js) = \int_{-\infty}^{+\infty} x_d(t)e^{-jst} \, dt = \int_{-\infty}^{+\infty} \sum_{n=-\infty}^{\infty} x(t)\delta(t - nT_s)e^{-jst} \, dt$$

$$= \sum_{n=-\infty}^{\infty} x(nT_s)e^{-jsnT_s},$$

where T_s is the sampling time interval. Assuming

$$x(n) \equiv x_d(nT_s) \equiv \{x(t)|_{t=nT_s}\} = \{x(nT_s)\}$$

we get the Laplace transform of the sequence $x(n)$ such that

$$X_d(js) = \sum_{n=-\infty}^{\infty} x(n)e^{-jnsT_s}.$$

Changing variable with $z = e^{js}$ and normalizing to $T_s = 1$, we obtain

$$X(z) = \sum_{n=-\infty}^{\infty} x(n)z^{-n}.$$

This is the definition of the z-transform of the sequence $x(n)$ which we will discuss in Chapter 3. Thus we can see that the Laplace transform of a discrete-time function produces a z-transform of the function by mapping the s-plane onto the z-plane. The mapping is done according to

$$z = e^{js},$$

where $s = \omega + j\delta$. Since

$$e^{js} = e^{j(\omega+j\delta)} = e^{j\omega-\delta} = e^{j(\omega+2\pi k)-\delta},$$

each of the frequency bands on the s-plane

$$2\pi(k-1) \leq \omega < 2\pi k \qquad k \text{ integer}$$

is mapped onto the z-plane. That is, the upper half and lower half of each band are mapped onto the interior and exterior regions of the unit circle on the z-plane, respectively. The real frequency axis is mapped onto the unit circle. This is shown in Fig. 2.3.8.

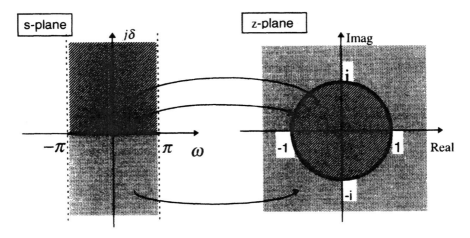

Figure 2.3.8   Correspondence between s-plane and z-plane.

2.4 Sampling Continuous Signals

2.4.1 Sampling and Interpolation

Consider a continuous-time signal $s(t)$ and a train of delta functions

$$p(t) = \sum_{n=-\infty}^{\infty} \delta(t - nT_s),$$

where n represents an integer and T_s is the sampling time interval. Since the sampling $s(t)$ can be interpreted as a multiplication $s(t)$ by $p(t)$ as shown in Fig. 2.4.1, we get a sampled function $s_d(t)$ as (refer to Section 2.3.1)

$$s_d(t) = s(t) \cdot p(t) = s(t) \sum_{n=-\infty}^{\infty} \delta(t - nT_s).$$

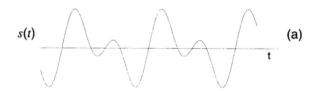

Figure 2.4.1 An illustration of sampling a continuous-time signal: (a) a continuous-time signal; (b) an impulse sequence; (c) the sample value sequence.

The function $p(t)$ is a periodic function composed of impulse functions with the period T_s, so $p(t)$ can be represented by a Fourier series expansion as

$$p(t) = \sum_{m=-\infty}^{\infty} c_m e^{j\omega_m t},$$

where

$$c_m = \frac{1}{T_s} \int_{-T_s/2}^{T_s/2} \delta(t) e^{-j\omega_m t} \, dt = \frac{1}{T_s}$$

and

$$\omega_m = \frac{2\pi}{T_s} m.$$

Therefore,

$$s_d(t) = s(t) \sum_{m=-\infty}^{\infty} c_m e^{j\omega_m t} = s(t) \frac{1}{T_s} \sum_{m=-\infty}^{\infty} e^{j\omega_m t}.$$

Let the Fourier transforms of $s_d(t)$ and $s(t)$ be $S_d(\omega)$ and $S(\omega)$, respectively. We show that $S(\omega)$ and $S_d(\omega)$ are equivalent under some condition. The Fourier transform $S_d(\omega)$ can be written as

$$S_d(\omega) = \int_{-\infty}^{\infty} s_d(t) e^{-j\omega t} \, dt = \int_{-\infty}^{\infty} s(t) \sum_{m=-\infty}^{\infty} c_m e^{j\omega_m t} \cdot e^{-j\omega t} \, dt$$

$$= \frac{1}{T_s} \sum_{m=-\infty}^{\infty} \int_{-\infty}^{\infty} s(t) e^{-j(\omega - \omega_m) t} \, dt = \frac{1}{T_s} \sum_{m=-\infty}^{\infty} S(\omega - \omega_m)$$

$$= \frac{1}{T_s} \sum_{m=-\infty}^{\infty} S\left(\omega - \frac{2\pi}{T_s} m\right),$$

where

$$S(\omega) = \int_{-\infty}^{\infty} s(t) e^{-j\omega t} \, dt.$$

This result shows that the Fourier transform (or spectrum density) $S_d(\omega)$ of the sampled function $s_d(t)$ is an infinite periodic sequence of the Fourier transform $S(\omega)$ of the original signal $s(t)$ separated by the period $2\pi/T_s$. This is shown in Fig. 2.4.2.

On the other hand $S_d(\omega)$ can be calculated using the sampled values $s(nT_s)$ (n integer) as

$$S_d(\omega) = \int_{-\infty}^{+\infty} s_d(t) e^{-j\omega t} \, dt = \int_{-\infty}^{\infty} s(t) \sum_{n=-\infty}^{\infty} \delta(t - nT_s) e^{-j\omega t} \, dt$$

$$= \sum_{n=-\infty}^{+\infty} \int_{-\infty}^{+\infty} s(t) \delta(t - nT_s) e^{-j\omega t} \, dt = \sum_{n=-\infty}^{\infty} s(nT_s) e^{-j\omega n T_s}.$$

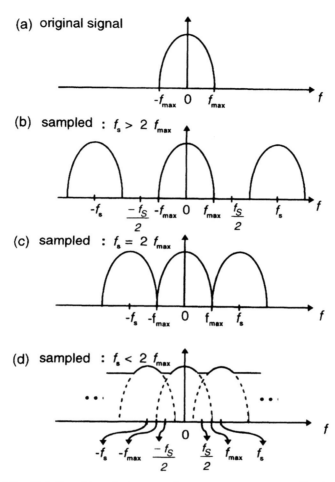

Figure 2.4.2 Relationship between frequency spectra of the original continuous-time signal with maximum frequency $f_{\max}$ and the sampled signal: (a) the original signal spectrum; (b) the sampled signal spectrum when the sampling frequency $f_s > 2f_{\max}$; (c) the sampled spectrum when $f_s = 2f_{\max}$; (d) the sampled signal spectrum when $f_s < 2f_{\max}$.

From this result and the relation obtained above,

$$S_d(\omega) = \frac{1}{T_s} \sum_{m=-\infty}^{+\infty} S\left(\omega - \frac{2\pi}{T_s}m\right),$$

we can relate $S_d(\omega)$ calculated using sample values and $S(\omega)$ as

$$S_d(\omega) = \frac{1}{T_s} S(\omega) = \sum_{n=-\infty}^{\infty} s(nT_s) e^{-j\omega nT_s},$$

where

$$-\frac{\pi}{T_s} \le \omega < \frac{\pi}{T_s}.$$

That is, if we restrict the frequency band as above, the Fourier transform of the discrete-time sequence $s_d(nT_s)$ is the same as the Fourier transform of the continuous-time signal $s(t)$ except for a constant $1/T_s$.

Next consider the waveform reconstruction of $s(t)$ from the sample values $s(nT_s)$. $S_d(\omega)$ is a periodic function of ω. So if we extract the one period of $S_d(\omega)$ which is the interval $\pi/T_s \le \omega < \pi/T_s$ by using an ideal low-pass filter, then the filter outcome $\hat{s}(t)$ is expected to be proportional to the original continuous-time signal $s(t)$. Let us use an ideal low-pass filter for the purpose such as

$$H_L(\omega) = \begin{cases} 1 & |\omega| < \pi/T_s \\ 0 & \text{otherwise.} \end{cases}$$

The output signal $\hat{s}(t)$ of the low-pass filter is shown to be identical to the original signal $s(t)$ multiplied by $1/T_s$ as

$$\hat{s}(t) = \frac{1}{2\pi} \int_{-\pi/T_s}^{\pi/T_s} H_L(\omega) \cdot S_d(\omega) e^{j\omega t}\, d\omega = \frac{1}{2\pi} \int_{-\pi/T_s}^{\pi/T_s} \frac{1}{T_s} S(\omega) \cdot e^{j\omega t}\, d\omega$$

$$= \frac{1}{T_s} s(t),$$

where we assume that

$$S_d(\omega) = \frac{1}{T_s} S(\omega) \qquad \text{for } |\omega| < \frac{\pi}{T_s}.$$

On the other hand, the signal $\hat{s}(t)$ is obtained from the sample values of $s(t)$ as

$$\hat{s}(t) = \frac{1}{2\pi} \int_{-\pi/T_s}^{\pi/T_s} S_d(\omega) e^{j\omega t}\, d\omega$$

$$= \frac{1}{2\pi} \int_{-\pi/T_s}^{\pi/T_s} \sum_{n=-\infty}^{+\infty} s(nT_s) e^{-j\omega nT_s} \cdot e^{j\omega t}\, d\omega$$

$$= \frac{1}{2\pi} \sum_{n=-\infty}^{+\infty} s(nT_s) \int_{-\pi/T_s}^{\pi/T_s} e^{j\omega(t-nT_s)}\, d\omega$$

$$= \frac{1}{T_s} \sum_{n=-\infty}^{+\infty} s(nT_s) \frac{\sin[(t-nT_s)\pi/T_s]}{(t-nT_s)\pi/T_s},$$

where

$$S_d(\omega) = \sum_{n=-\infty}^{+\infty} s(nT_s)e^{-j\omega nT_s}$$

and

$$\frac{1}{2\pi} \int_{-\pi/T_s}^{\pi/T_s} e^{j(t-nT_s)\omega} \, d\omega = \frac{1}{T_s} \frac{\sin[(t-nT_s)\pi/T_s]}{(t-nT_s)\pi/T_s}.$$

Remembering that

$$\hat{s}(t) = \frac{1}{T_s} s(t)$$

and that the sample values of $s(t)$ at $t = nT_s$ are given by $s(nT_s)$, the signal values at time other than $t = nT_s$ are determined by convolution with the interpolation function as

$$s(t) = \sum_{n=-\infty}^{+\infty} s(nT_s) \frac{\sin(t-nT_s)\pi/T_s}{(t-nT_s)\pi/T_s}.$$

Therefore, the continuous-time signal $s(t)$ can be completely reconstructed from the sample values of $s(t)$. Figure 2.4.3 illustrates an example of interpolation of several sample values.

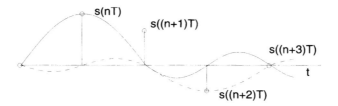

Figure 2.4.3 An illustration of the interpolation of several sample values where the signal values at the time between $t = nT$ and $(n+1)T$ are obtained by summing all the interpolation function values at t.

In the discussions above, we assumed that the Fourier spectrum density $S(\omega)$ of the original signal $s(t)$ is band-limited to the maximum frequency $\omega_{\max} < \pi/T_s$. This assumption allows us to extract $S(\omega)$ components from the $S_d(\omega)$ by ideal low-pass filtering. If $\omega_{\max} < \pi/T_s$ is not satisfied, we cannot extract the correct $S(\omega)$ components, since the adjacent periods of $S(\omega)$ overlap each other as shown in Fig. 2.4.2(d). This condition is called *aliasing* and will be discussed in more detail in the next section.

2.4.2   Aliasing and Sampling Theorem

Again write the Fourier transform of the sampled function $s_d(t)$ as

$$S_d(\omega) = \frac{1}{T_s} \sum_{m=-\infty}^{\infty} S\left(\omega - \frac{2\pi}{T_s} m\right),$$

where $S_d(\omega)$ is called the Fourier spectrum density of the sample value sequence and $S(\omega)$ is the spectrum density of the original continuous-time signal. Let $S(\omega)$ be the band-limited spectrum to the maximum frequency ω_{max}. $S_d(\omega)$ is periodic with the period $2\pi/T_s$. If $\omega_{max} < \pi/T_s$, then

$$S_d(\omega) = \frac{1}{T_s} S(\omega) \qquad \left(-\frac{\pi}{T_s} \leq \omega \leq \frac{\pi}{T_s}\right)$$

holds and the $S(\omega)$ can be extracted without any distortion from the $S_d(\omega)$ using an ideal low-pass filter with the pass-band

$$-\frac{\pi}{T_s} \leq \omega \leq \frac{\pi}{T_s}.$$

If $\omega_{max} \geq \pi/T_s$, however,

$$S_d(\omega) \neq \frac{1}{T_s} S(\omega)$$

because $S_d(\omega \pm \pi/T_s)$ and $S(\omega)$ overlap in the frequency region as $\pi/T_s \leq |\omega| \leq \omega_{max}$; thus $S(\omega)$ will be distorted. This spectral distortion is called the *aliasing noise* or error. The aliasing noise affects the reconstructed signal waveform. To avoid the aliasing noise effects the condition

$$\omega_{max} < \frac{\pi}{T_s}$$

must hold, or equivalently,

$$f_{max} < \frac{1}{2T_s} \equiv \frac{f_s}{2}.$$

Then we impose

$$T_s < \frac{1}{2f_{max}},$$

where $\omega_{max} = 2\pi f_{max}$. Thus we have the sampling theorem:

Sampling Theorem   *A band-limited signal waveform with the maximum frequency f_{max} is reconstructed from the sample values sampled at a* *uniform time interval shorter than the inverse of* $2f_{\max}$. *The waveform reconstruction is performed according to*

$$s(t) = \sum_{n=-\infty}^{+\infty} s(nT_s)\frac{\sin\{(t-nT)\pi/T_s\}}{(t-nT_s)\pi/T_s}.$$

Example 2.4.1

Consider an exponential continuous-time function $f(t)$ as

$$f(t) = \begin{cases} e^{-at} & t \geq 0 \\ 0 & t < 0, \end{cases}$$

where a is a real positive constant. The Fourier transform $F(\omega)$ is given by

$$F(\omega) = \int_0^\infty e^{-at}e^{-j\omega t}\,dt = \int_0^\infty e^{-(a+j\omega t)}\,dt = \left[-\frac{1}{a+j\omega}e^{-(a+j\omega)t}\right]_0^\infty$$

$$= \frac{1}{a+j\omega}.$$

The magnitude spectrum $F(\omega)$ is

$$|F(\omega)| = \sqrt{\frac{1}{a^2+\omega^2}},$$

where $f(t)$ and $|F(\omega)|$ are illustrated in Fig. 2.4.4(a) and (b). The maximum frequency of $F(\omega)$ is not clear in the figure, so we suppose the maximum frequency to be $\omega_{\max}$, satisfying the relation

$$\left|\frac{F(\omega_{\max})}{F(0)}\right|^2 = \frac{1}{2}.$$

Figure 2.4.4(c) and (d) show the frequency spectra $F_d(\omega)$ obtained by sampling $f(t)$ with the sampling frequencies $2\omega_{\max} = \omega_{s1}$ and $4\omega_{\max} = 2\omega_{s1} = \omega_{s2}$, respectively. We can see the effects of aliasing on the spectra of the original time function.

2.5 Summary

In this chapter we first described the basic issues for processing continuous-time signals in a digital processor implemented with computer software and for hardware. Discrete expressions for signals in the time domain were presented. The input–output relations of linear time-invariant

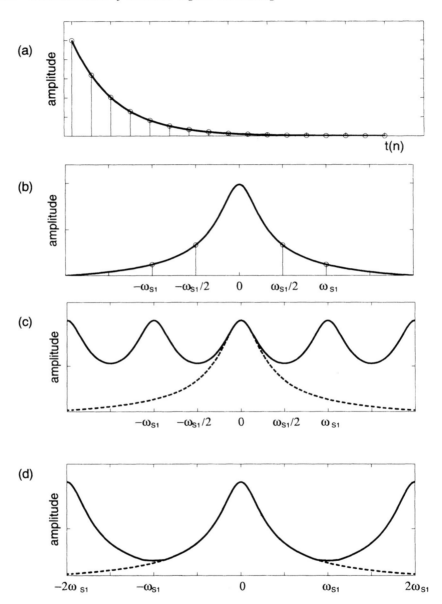

Figure 2.4.4   An example of aliasing arising in the frequency spectrum of an exponential time function $f(t) = e^{-at}$, where a is a real constant less than unity: (a) sampled sequence of the time function $f(t)$ shown by the solid line; (b) Fourier transform $F(\omega)$ of $f(t)$; (c) and (d) Fourier spectra when the sampling frequencies are ω_{s1} and $\omega_{s2}(= 2\omega_{s1})$, respectively, where the frequency bandwidth of $F(\omega)$ is determined by the frequency which gives the $-6\,\mathrm{dB}$ fall off of $F(\omega)$.

discrete systems were shown using a unit impulse response (impulse response) sequence and we obtained a convolution operation between an input sequence and the impulse response sequence. When a system is described by a constant coefficient linear difference equation, we presented the basic structure of the system implemented using delay elements, multipliers, and adders. We reviewed briefly the Fourier series expansion of a periodic sequence and the Fourier transform of a transient sequence. We also presented a brief discussion on the Laplace transform. Finally, the sampling theorem was presented for continuous-time functions together with the exact reconstruction of the original functions using an ideal low-pass filter.

3 z-Transform

3.0   Introduction

The z-transform is very useful in the discrete-time world and plays an important role in system design and analysis. The Fourier transform of a sequence is obtained by evaluating the z-transform on the unit circle which has unit radius centered at $z = 0$. The unit circle on the z-plane is denoted by

$$z = e^{j\omega},$$

where ω is a normalized angular frequency the range of which is

$$-\pi \leq \omega < \pi.$$

Evaluating the z-transform function, say $F(z)$, on the unit circle is to substitute $z = e^{j\omega}$ into $F(z)$. We get the Fourier transform as

$$F(e^{j\omega}) = F(z)|_{z=e^{j\omega}}.$$

We see that the Fourier transform of a sequence becomes a periodic function of ω with a period of 2π. Since practical sequences or systems are described by their Fourier transforms, it is very important to recognize that the Fourier transform can be obtained by evaluating the z-transform on the unit circle. Analysis and design for sequences or systems can be carried out on the entire z-plane and as a result, when sequences or systems are physically realizable, replacing z with $e^{j\omega}$ leads to the characteristics of the sequences and systems which we want to know.

It is not possible, however, to know the physical realizability of sequences or systems from the Fourier transforms only. The complex variable s is used in the Laplace transform of continuous-time signals as described in Section 2.3.4. The variable s is an extension of the real frequency to the complex frequency by the relation

$$s = \omega + j\delta,$$

where δ is a real variable corresponding to the loss factor of a system or the decay factor of a signal and ω is a real variable expressing the real angular frequency.

The complex variable z is customarily related to s by

$$z = e^{jsT_s},$$

where T_s is the time interval used for sampling continuous-time signals. The definition $z = e^{jsT_s}$ maps any real frequency band in the s-plane limited by the intervals $-\pi + 2k\pi \leq \omega < \pi + 2k\pi$ (k integer) onto the unit circle in the z-plane. The upper half of any band on the s-plane is mapped onto the inner part, while the lower half part is mapped onto the outer region of the unit circle, as shown in Fig. 2.3.8.

3.1 Definition of the z-Transform

The z-transform $S(z)$ of a sequence $s(n)$ is defined as

$$S(z) = \sum_{n=-\infty}^{\infty} s(n)z^{-n},$$

where z is a complex variable. We assume that z takes any point on the entire z-plane and ∞. This definition is called the *two-sided* z-transform, because it contains the sequence values for all n. Another definition which does not take the sample values for $n < 0$ into account is called the *one-sided* z-transform. When the sample values of the sequence $s(n)$ for $n < 0$ are all zero, the two definitions give us an identical z-transform.

The z-transform defined above is an infinite power series, but the series does not converge under some conditions. It is known that the sufficient condition for the power series to converge is that the series is absolutely summable:

$$\sum_{n=-\infty}^{\infty} |s(n)z^{-n}| < \infty.$$

The region of the z-plane where the power series converges is referred to as a region of convergence. The region is the inside or the outside of a circle of radius R, or an annular region between two circles of different radii R_1 and R_2 on the z-plane. That is, the convergence region can be described in terms of the properties of the sequence such as

$$|z| < R,$$

$$|z| > R,$$

or

$$R_1 < |z| < R_2,$$

where R, R_1 and R_2 are real numbers larger than 0. Examples of these regions of convergence are shown in Fig. 3.1.1. A z-transform $S(z)$ has no singularity (pole) in the region of convergence and we say that $S(z)$ is analytic in the region.

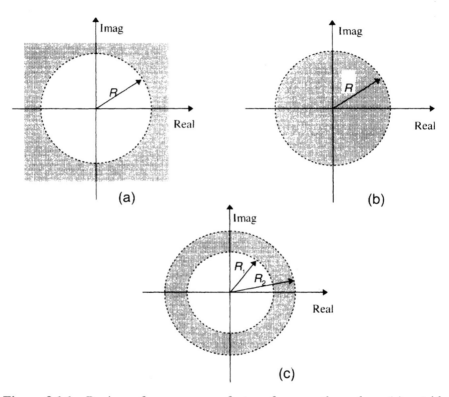

Figure 3.1.1 Regions of convergence of z-transforms on the z-plane: (a) outside the circle; (b) insides the circle; (c) annular region formed by the intersection of two circles.

Example 3.1.1

Consider an exponential series $s_1(n)$ defined by

$$s_1(n) = \begin{cases} a_1^n & n \geq 0 \\ 0 & \text{otherwise,} \end{cases}$$

where $a_1 = $ constant. The z-transform $S_1(z)$ of the series following the definition is

$$S_1(z) = \sum_{n=-\infty}^{\infty} s_1(n)z^{-n} = \sum_{n=0}^{\infty}(a_1z^{-1})^n = \frac{1}{1-a_1z^{-1}},$$

where $|a_1z^{-1}| < 1$. This tells us that if $|z| > |a_1|$, $S_1(z)$ converges and consequently the region of convergence is $|z| > |a_1|$, which is the outer part of a circular region with radius $|a_1|$ centered at the origin of the z-plane. The sequence $s_1(n)$ and the region of convergence of $S_1(z)$ are illustrated in Fig. 3.1.2 for $a_1 = 0.7$.

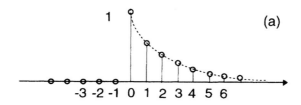

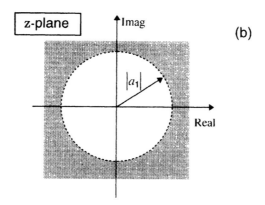

Figure 3.1.2 The region of convergence of an exponential sequence $s_1(n)$:
(a) $s_1(n) = a_1^n$ ($n \geq 0$); (b) region of convergence $|z| > |a_1|$.

Example 3.1.2

Consider a sequence $s_2(n)$ of the form

$$s_2(n) = \begin{cases} -a_2^n & \text{for } n < 0 \\ 0 & \text{otherwise,} \end{cases}$$

where a_2 is a constant. The z-transform is calculated with the result

$$S_2(z) = -\sum_{n=-\infty}^{-1} a_2^n z^{-n} = \frac{1}{1-a_2z^{-1}},$$

where $|a_2 z^{-1}| > 1$. We see that the z-transform of an exponential sequence with nonzero values in the range $-\infty < n < 0$ will have the region of convergence $|z| < |a_2|$ which is the inner region of a circle centered at $z = 0$ with radius $|a_2|$. The sequence $s_2(n)$ and the region of convergence of $S_2(z)$ for real a_2 are illustrated in Fig. 3.1.3.

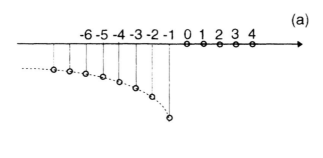

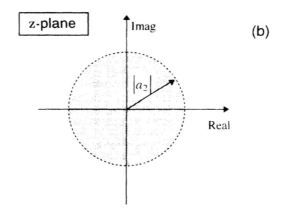

Figure 3.1.3   The region of convergence of an exponential sequence $s_2(n)$:
(a) $s_2(n) = -a_2^n$ $(n < 0)$; (b) region of convergence $|z| < |a_2|$.

Compare the two z-transforms $S_1(z)$ and $S_2(z)$. If $a_1 = a_2$, they are identical in formal expression but their regions of convergence are different. In fact the regions of convergence of Examples 3.1.1 and 3.1.2 are

$$|z| > |a_1| \qquad \text{for } S_1(z)$$

and

$$|z| < |a_2| \qquad \text{for } S_2(z),$$

respectively. From this, a z-transform should be defined together with its region of convergence. In other words, a time sequence cannot be uniquely defined by a z-transform without indication of the region of convergence.

Example 3.1.3

Consider the sequence given by the sum of the sequences $s_1(n)$ and $s_2(n)$ which are discussed in Examples 3.1.1 and 3.1.2:

$$s_3(n) = s_1(n) + s_2(n).$$

The sequence is a series having nonzero values in the range $-\infty < n < \infty$. The z-transform $S_3(z)$ is

$$S_3(z) = S_1(z) + S_2(z) = \frac{1}{1 - a_1 z^{-1}} + \frac{1}{1 - a_2 z^{-1}}.$$

The region of convergence must be the intersection of the two regions of convergence. Since $S_1(z)$ and $S_2(z)$ have regions of convergence $|z| > |a_1|$ and $|z| < |a_2|$, respectively, the region of convergence of $S_3(z)$ is as follows:

    (i)   if $|a_1| < |a_2|$, then $|a_1| < |z| < |a_2|$

or

    (ii)   if $|a_1| > |a_2|$, then there is no region of convergence.

    The sequence $s_3(n)$ and the region of convergence are depicted in Fig. 3.1.4 for $a_1 = 0.7$ and $a_2 = 0.9$.

3.2   The Regions of Convergence for Different Kinds of Sequences

Referring to the regions of convergence discussed in Examples 3.1.1 through 3.1.3, consider the regions of convergence of general sequences. There are four kinds of sequences to be considered, provided that N, N_1, and N_2 are integers.

    (i)   A class of sequences with nonzero value for $n \geq N$. A sequence contained in this class is called a *right-sided* sequence and is of the form

$$s(n) = \begin{cases} s(n) & n \geq N \\ 0 & n < N. \end{cases}$$

    (ii) A class of sequences with non-zero values for $n < N$. A sequence of this class is called a *left-sided* sequence and defined as

$$s(n) = \begin{cases} s(n) & \text{for } n \leq N \\ 0 & \text{for } n > N. \end{cases}$$

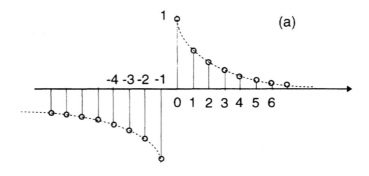

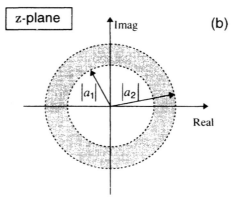

Figure 3.1.4   The region of convergence of $s(n) = s_1(n) + s_2(n)$ when $a_1 = 0.7$ and $a_2 = 0.9$: (a) $s(n)$; (b) region of convergence $|a_1| < |z| < |a_2|$.

(iii)   A class of sequences with nonzero values for a finite interval $N_1 \le n \le N_2$. A sequence in this class is called a *finite-length sequence* and is expressed as

$$s(n) = \begin{cases} s(n) & \text{for } N_1 \le n \le N_2 \\ 0 & \text{otherwise.} \end{cases}$$

(iv)   A class of sequences with nonzero values for all n. A sequence in this class is expressed as

$$s(n) = s(n) \qquad \text{for all } n.$$

3.2.1   Finite Sequences

Let us consider first the region of convergence of a sequence in class (iii). The z-transform of a sequence in this class is defined by

$$S(z) = \sum_{n=N_1}^{N_2} s(n)z^{-n} \qquad N_2 > N_1.$$

Since $S(z)$ is a sum of finite terms, $S(z)$ exists on the entire z-plane except at $z = 0$ and/or $z = \infty$. Since z^{-n} becomes infinity depending on n and z, if $N_2 > N_1 \geq 0$, $z = 0$ is not contained in the region of convergence. If $N_2 > 0 > N_1$, $z = 0$ and $z = \infty$ are not included. If $N_1 < N_2 < 0$, $z = \infty$ is not contained in the convergent region.

3.2.2 Right-sided Sequences

Next consider the region of convergence of a sequence in class (i), i.e., right-sided sequences. If $N < 0$, then the z-transform becomes

$$S(z) = \sum_{n=N}^{\infty} s(n)z^{-n}.$$

The partial sum of $S(z)$,

$$\sum_{n=N}^{-1} s(n)z^{-n},$$

exists on the entire z-plane, because this is a sum of a finite number of terms. The remaining series of the sequence,

$$\sum_{n=0}^{\infty} s(n)z^{-n},$$

converge if the series is absolutely summable; that is, if

$$\sum_{n=0}^{\infty} |s(n)z^{-n}| < \infty.$$

We assume that $S(z)$ converges at z on a circle with a radius R_0; that is, we can find a real number R_0 which satisfies

$$\sum_{n=N}^{\infty} s(n)R_0^{-n} < \infty.$$

For any z on a circle with an arbitrary radius R larger than R_0,

$$|z| \geq |Re^{j\theta}| > R_0,$$

we get

$$\sum_{n=0}^{\infty} |s(n)z^{-n}| \leq \sum_{n=0}^{\infty} |s(n)R^{-n}| < \sum_{n=0}^{\infty} |s(n)R_0^{-n}| < \infty.$$

This indicates that the z-transform of the sequence $s(n)$ exists in the region

$$|z| \geq R > R_0,$$

where R is an arbitrary number larger than R_0. Therefore, the region of convergence of any right-sided sequence is the outer region of a circle centered at $z = 0$. The radius of the circle depends on the sequence. We could have a little more detailed discussion on the radius of the circle in the case when the z-transform of a sequence is expressed by a rational function of z^{-1} as in Examples 3.1.1 and 3.1.2.

A rational function is defined by a ratio of polynomials. The zeros of the denominator polynomial are the poles of the rational function. Recall Example 3.1.1 in which we have the z-transform $S_1(z)$ of the right-sided sequence $s_1(n) = a_1^n (n \geq 0)$ expressed by

$$S_1(z) = \frac{1}{1 - a_1 z^{-1}} = \frac{z}{z - a_1} \qquad |z| > |a_1|.$$

This is an example of a rational function whose pole is $z = a_1$ and whose region of convergence is $|z| > |a_1|$. This suggests that the pole locations of the rational function will determine the region of convergence of the z-transform. In fact it can be proved that the region of convergence for a rational function type of z-transform is the outer region of a circle centered at $z = 0$ with a radius which is equal to the largest magnitude of the poles.

If the poles of a rational function are $P_1, P_2, \ldots, P_L$, then the region of convergence is

$$|z| > |P_{\max}| = \max(|P_1|, |P_2|, \ldots, |P_L|).$$

As will be discussed in Section 3.3, any rational function can be expanded into a sum of partial fractions each of which is of the form $1/(1 - P_i z^{-1})$ and which has a region of convergence given by $|z| > P_i$, if a right-sided sequence is assumed. Therefore, the region of convergence of the rational function must be the intersection of the regions of convergence for all the partial fractions. This results in the above region of convergence, which is illustrated in Fig. 3.2.1 when the number of poles (L) is 3.

In practical applications we often encounter z-transforms described by rational functions of z. If we know $|P_{\max}| < 1$, then the region of convergence includes the unit circle and therefore the Fourier transform exists as long as we deal with right-sided sequences.

3.2.3 Left-sided Sequences

The region of convergence for a left-sided sequence can be investigated in a similar manner to the discussions on the right-sided sequences in the

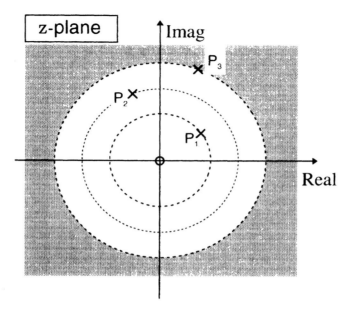

Figure 3.2.1 Region of convergence of the z-transform of a right-sided sequence, where P_3 is the outermost pole in the P_1–P_3 and the region of convergence is $|z| > |P_3|$.

previous section. The z-transform of sequence $s(n)$ in class (ii) as defined in Section 3.2 is described by

$$S(z) = \sum_{n=-\infty}^{N} s(n)z^{-n}.$$

If $N > 0$, then $S(z)$ can be divided into two partial sums:

$$S(z) = \sum_{n=-\infty}^{0} s(n)z^{-n} + \sum_{n=1}^{N} s(n)z^{-n}.$$

In this equation the second term on the right-hand side is the z-transform of a finite-length sequence, so it exists on the entire z-plane except at the origin $z = 0$. Then it is sufficient to consider the convergence region of the first term,

$$\sum_{n=-\infty}^{0} s(n)z^{-n} = A(z),$$

to obtain the region of convergence of $S(z)$.

Assume that $A(z)$ converges at z on a circle

$$|z| = |R_0 e^{j\theta}| = R_0$$

with a radius R_0. Then we get

$$|A(z)| \leq \sum_{n=-\infty}^{0} |s(n)z^{-n}| \leq \sum_{n=-\infty}^{0} |s(n)R_0^{-n}| < \infty$$

for z on the circle $|z| = R_0$. If we take an arbitrary radius R satisfying

$$R < R_0,$$

then we get

$$\sum_{n=-\infty}^{0} |s(n)z^{-n}| \leq \sum_{n=-\infty}^{0} |s(n)R^{-n}| < \sum_{n=-\infty}^{0} |s(n)R_0^{-n}| < \infty$$

for z in the region $|z| \leq R$. From the discussion above we can see that the region of convergence of a left-sided sequence for $N > 0$ is the inner region of a circle excluding $z = 0$.

If $N \leq 0$, $S(z)$ becomes

$$S(z) = \sum_{n=-\infty}^{N} s(n)z^{-n}.$$

Since

$$|S(z)| \leq \sum_{n=-\infty}^{N} |s(n)z^{-n}| \leq \sum_{n=-\infty}^{0} |s(n)z^{-n}| < \infty,$$

we get that the region of convergence of $S(z)$ is the inner region of a circle in the same way as for the case $N > 0$.

Let us determine the radius R when the z-transform of a left-sided sequence $s(n)$ is expressed by a rational function $H(z)$. As discussed in the previous section, a rational function can be expanded into a sum of partial fractions each of which has the form

$$\frac{1}{1 - P_i z^{-1}},$$

where P_i is a zero of the denominator of the rational function and is a pole of the rational function.

Since the sequence $s(n)$ is left-sided, the region of convergence of each partial fraction is obtained from the result of Example 3.1.2 as

$$|P_i z^{-1}| > 1,$$

or equivalently,

$$|z| < |P_i|.$$

If $H(z)$ has k poles

$$P_1, P_2, \ldots, P_k,$$

then the region of convergence of $H(z)$ is obtained by the intersection of all the regions of convergence of all the partial fractions. Thus we have that the region of convergence of $H(z)$ is

$$|z| < R = |P_{\min}| = \min(|P_1|, |P_2|, \ldots, |P_k|).$$

This indicates that the region of convergence of a rational function derived for a left-sided sequence is the inner region of a circle whose radius is equal to the magnitude of the smallest-magnitude pole.

3.2.4 Two-sided Sequences

The region of convergence of a two-sided sequence can be discussed in the same way as in the previous two sections. A two-sided sequence $s(n)$ is divided into two parts,

$$s(n) = s_l(n) + s_r(n),$$

where $s_l(n)$ has nonzero values for $-\infty < n < 0$ and $s_r(n)$ has nonzero values for $0 \leq n < \infty$. Thus $s_l(n)$ is a left-sided sequence and $s_r(n)$ is a right-sided sequence. The z-transform of $s(n)$ is

$$S(z) = \sum_{n=-\infty}^{\infty} s(n)z^{-n} = \sum_{n-\infty}^{-1} s_l(n)z^{-n} + \sum_{n=0}^{\infty} s_r(n)z^{-n}$$

$$= S_l(z) + S_r(z).$$

The first term $S_l(z)$ has been discussed on the region of convergence for left-sided sequences in Section 3.2.3 and the second term $S_r(z)$ has also been discussed for right-sided sequences in Section 3.2.2. The regions of convergence of $S_l(z)$ and $S_r(z)$ are expressed by

$$|z| < |a|$$

and

$$|z| > |b|,$$

respectively, where $|a|$ is equal to the magnitude of the smallest-magnitude pole of $S_l(z)$ and $|b|$ equals the magnitude of the largest-magnitude pole of $S_r(z)$.

Accordingly, the region of convergence of $S(z)$ is obtained by the intersection of the two regions. If $|a| < |b|$, then the intersection does not exist and therefore $S(z)$ has no region of convergence. On the contrary, if $|a| > |b|$, then we have a common annular region as the region of convergence where there is no pole of $S(z)$. If the annular region contains the unit circle, the two-sided sequence has a Fourier transform which is expressed by

$$S(e^{j\omega}) = \sum_{n=-\infty}^{\infty} s(n)e^{-j\omega n}.$$

3.2.5 Examples of z-Transform Pairs

In the previous section we defined the z-transform of a sequence and illustrated z-transforms of some simple but fundamental sequences. It was pointed out that z-transform functions must be defined with their regions of convergence. In this section a couple of important sequences and their z-transforms, called z-transform pairs, are summarized in Table 3.2.1. Table 3.2.1 includes only a few sequences and corresponding z-transforms; however in many practical applications, particularly for dealing with linear and time-invariant systems, the characteristics of which are expressed by a constant coefficient difference equation, the table is useful for finding a sequence from $H(z)$ given by a rational function. This process is called inverse z-transform and will be described in Section 3.3.

Some sequences and corresponding z-transforms listed in Table 3.2.1 have not been discussed in the preceding sections, but the z-transforms will be obtained directly by mathematical manipulations.

3.3 Inverse z-Transform

It is often necessary to get a time sequence from a given z-transform. For example the z-transform of a filter designed on the z-plane should be converted to a time sequence which gives an impulse response sequence for implementation by a hardware or software system.

Another example is the case of two systems with cascade connection. The impulse response sequence of the total system can be obtained by multiplying the two z-transforms and in turn performing the inverse z-transform of the product. Although there is another way to calculate the total impulse response sequence by forming the convolution sum of the

Table 3.2.1 Examples of z-transform pairs.

$u(n) = 1$ for $n \geq 0$
$\quad\quad = 0$ otherwise

Sequence		z-Transform	Region of Convergence	Comments
m-sample delayed unit impulse	$\delta(n-m)$	z^{-m}	$m > 0 \to$ all z except $z = 0$ $m = 0 \to$ all z $m < 0 \to$ all z except $z = \infty$	m: integer constant
unit impulse sequence	$u(n)$	$\dfrac{1}{1-z^{-1}}$	$|z| < 1$	
	$-u(-n-1)$		$|z| > 1$	
exponential sequence	$a^n u(n)$	$\dfrac{1}{1-az^{-1}}$ $\quad(*)$	$|z| < |a|$	a: complex constant
	$-a^n u(-n-1)$		$|z| > |a|$	
damped complex sinusoidal sequence	$r^n e^{j\omega_0 n} u(n)$	$\dfrac{1}{1-re^{j\omega_0}z^{-1}}$	$|z| > r$	
	$-r^n e^{j\omega_0 n} u(-n-1)$		$|z| < r$	
exponential sequence multiplied by time	$na^n u(n)$	$\dfrac{az^{-1}}{(1-az^{-1})^2}$	$|z| > |a|$	differentiation with respect to z of z-transform indicated by $(*)$
	$-na^n u(-n-1)$		$|z| < |a|$	

two sequences, the multiplication method is convenient, especially when the inverse z-transform of the total system can be calculated efficiently.

3.3.1   Inverse z-Transform Using Contour Integral

The inverse z-transform of $H(z)$ is defined using the contour integral, referring to complex function theory, as follows:

$$h(n) = \frac{1}{2\pi j} \oint_c H(z) z^{n-1}\, dz,$$

where $h(n)$ is a time sequence, c is the contour along which the integral is taken is a closed curve located in the region of convergence of $H(z)$, and $z = 0$ is included in the inner region surrounded by the contour c.

Let us confirm the inverse z-transform formula. From the definition of the z-transform, it follows that $H(z)$ is expressed by

$$H(z) = \sum_{n=-\infty}^{\infty} h(n) z^{-n}.$$

Substitution of this $H(z)$ into the inverse z-transform definition above leads to

$$\frac{1}{2\pi j} \oint_c H(z) z^{k-1}\, dz = \frac{1}{2\pi j} \oint_c \left\{ \sum_{n=-\infty}^{\infty} h(n) z^{-n} \right\} z^{k-1}\, dz$$

$$= \frac{1}{2\pi j} \sum_{n=-\infty}^{\infty} h(n) \oint_c z^{-n+k-1}\, dz.$$

Here we show Cauchy's theorem, which tells us that the following contour integral holds:

$$\frac{1}{2\pi j} \oint_c z^{n-1}\, dz = \begin{cases} 1 & \text{for } n = 0 \\ 0 & \text{for } n \neq 0, \end{cases}$$

where c is a closed curve encircling the origin $z = 0$ and the integration is carried out anticlockwise along c. Cauchy's theorem is confirmed by choosing the closed curve as a circle with the radius R and centered at $z = 0$ as shown in Fig. 3.3.1.

The circle can be expressed in polar coordinates as

$$z = Re^{j\theta} \qquad 0 \leq \theta \leq 2\pi.$$

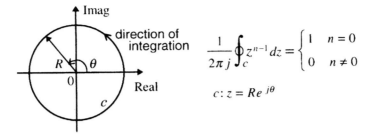

Figure 3.3.1 Cauchy's theorem. The contour integral is performed along c, which is a closed curve centered at $z = 0$. In this example, c is a circle of radius R.

The contour integral I,

$$I = \frac{1}{2\pi j} \oint_c z^{n-1}\, dz,$$

is changed to a line integral

$$I = \frac{1}{2\pi j} \oint_c z^{n-1}\, dz = \frac{1}{2\pi j} \oint_c (Re^{j\theta})^{n-1} R e^{j\theta}\, d\theta = \frac{1}{2\pi} \int_0^{2\pi} R^n e^{j\theta n}\, d\theta,$$

where we have used the relation

$$dz = jRe^{j\theta}\, d\theta.$$

Calculating the line integral we get

$$I = \frac{1}{2\pi} \int_0^{2\pi} R^n e^{j\theta n}\, d\theta = \frac{1}{2\pi} \frac{R^n}{jn} e^{j\theta n}\Big|_{\theta=0}^{2\pi} = 0 \qquad \text{for } n \neq 0$$

and

$$I = \frac{1}{2\pi} \int_0^{2\pi} d\theta = 1 \qquad \text{for } n = 0.$$

This is the result we want.

Using Cauchy's theorem, the inverse z-transform formula is reduced to

$$\frac{1}{2\pi j} \oint_c H(z) z^{k-1}\, dz = \frac{1}{2\pi j} \sum_{n=-\infty}^{\infty} h(n) \oint_c z^{-n+k-1}\, dz$$

$$= \begin{cases} h(k) & \text{for } n = k\ (-n+k=0) \\ 0 & \text{for } n \neq k\ (-n+k \neq 0). \end{cases}$$

From the discussion above we can see that the inverse z-transform of $H(z)$ is obtained by

$$h(n) = \frac{1}{2\pi j} \oint_c H(z) z^{n-1}\, dz,$$

where the variable k is substituted by n.

Although this contour integral can easily be evaluated for a z-transform expressed as a power series in z, it is not always easy to calculate the contour integral when $H(z)$ has a closed form. In that case complex function theory has a useful theorem which is called the residue theorem. According to the residue theorem, the contour integral along c of a function $X(z)$ is obtained by the sum of the residues included in the integral loop c,

$$\frac{1}{2\pi j} \oint_c X(z)\, dz = \sum_{i=1}^{k} \mathrm{Res}[X(z)|z = z_i],$$

where the loop c is contained in region of convergence of $X(z)$ and encircles all the k poles $(z_1, z_2, \ldots, z_k)$ of $X(z)$ as illustrated in Fig. 3.3.2. $\mathrm{Res}[X(z)|z = z_i]$ denotes the residue of $X(z)$ at $z = z_i$.

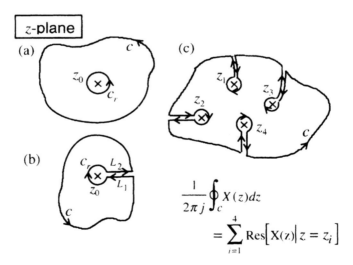

Figure 3.3.2   Examples of integral contours for the residue theorem.

The residue is calculated as

$$\mathrm{Res}[X(z)|z = z_i] \equiv (z - z_i)X(z)|_{z=z_i}$$

if z_i is a first-order pole. When z_i is an *m*th order $(m > 1)$ pole, the residue calculation is modified to

$$\text{Res}[X(z)|z = z_i] \equiv \frac{1}{(m-1)!} \left[\frac{d^{m-1}}{dz^{m-1}} (z - z_i)^m X(z) \right] \Bigg|_{z=z_i}.$$

A brief proof of the residue theorem is as follows.

Consider the contour integral of a function $X(z)$ along the closed loop c. We assume that $X(z)$ has an *m*th order pole z_0 inside c, as shown in Fig. 3.3.2(a). Using Laurent's expansion at the neighborhood of z_0, $X(z)$ can be expanded as

$$X(z) = \cdots + C_2(z - z_0)^2 + C_1(z - z_0) + C_0 + C_{-1}\frac{1}{(z - z_0)} + \cdots$$

$$+ C_{-m}\frac{1}{(z - z_0)^m}$$

$$= \sum_{k=-\infty}^{m} C_{-k}(z - z_0)^{-k}.$$

If we make a small circle c_r with radius r centered at $z = z_0$, as shown in Fig. 3.3.2(a) the contour integral of $X(z)$ along c_r is obtained as follows. Using the property that

$$\frac{1}{2\pi j} \oint_{c_r} C_{-k}(z - z_0)^{-k} \, dz = \frac{1}{2\pi j} \int_0^{2\pi} \frac{C_{-k}}{(re^{j\theta})^k} jre^{j\theta} \, d\theta$$

$$= \frac{C_{-k}}{2\pi r^{k-1}} \int_0^{2\pi} e^{-j(k-1)\theta} \, d\theta$$

$$= \begin{cases} C_{-1} & \text{for } k = 1 \\ 0 & \text{otherwise,} \end{cases}$$

we get

$$\frac{1}{2\pi j} \oint_{c_r} X(z) \, dz = C_{-1}.$$

If we build two paths L_1 and L_2 between c and c_r as shown in Fig. 3.3.2(b), the contour integral along the closed path p $(c \to L_1 \to c_r \to L_2)$ becomes 0, since no poles are inside the closed path p. Therefore, we have

$$I = \frac{1}{2\pi j} \oint_p X(z) \, dz$$

$$= \frac{1}{2\pi j} \left\{ \int_c X(z)\,dz + \int_{L_1} X(z)\,dz - \int_{c_r} X(z)\,dz + \int_{L_2} X(z)\,dz \right\}$$

$$= 0.$$

Since the paths L_1 and L_2 are the same path but of opposite directions, we have

$$\frac{1}{2\pi j} \int_{L_1} X(z)\,dz + \frac{1}{2\pi j} \int_{L_2} X(z)\,dz = 0.$$

Note that the contour integral of $X(z)$ along c_r is performed clockwise. This accounts for the negative sign before the contour integral along c_r in the equation for I. From the fact that $I = 0$, we get

$$\frac{1}{2\pi j} \int_c X(z)\,dz = \frac{1}{2\pi j} \int_{c_r} X(z)\,dz = C_{-1},$$

where C_{-1} is called the residue of $X(z)$ at $z = z_0$ and is expressed as

$$\mathrm{Res}[X(z)|z = z_0].$$

If the closed loop p contains N poles $(z_1, z_2, \ldots, z_N)$ of $X(z)$, then we have

$$\frac{1}{2\pi j} \oint_p X(z)\,dz = \sum_{i=1}^{N} \mathrm{Res}[X(z)|z = z_i].$$

Next we show the residue calculation method for $X(z)$ at $z = z_0$. According to Cauchy's integral theorem, an analytic function $f(z)$ is expressed by a contour integral as

$$f(z) = \frac{1}{2\pi j} \oint_{c\hat{r}} \frac{f(\zeta)}{\zeta - z}\,d\zeta,$$

where $c\hat{r}$ is a small circle with radius r centered at $\zeta = z$ and $f(z)$ is analytic on and inside the loop $c\hat{r}$. The derivative of $f(z)$ is given by

$$\frac{df(z)}{dz} = \frac{1}{2\pi j} \oint_{c\hat{r}} \frac{f(\zeta)}{(\zeta - z)^2}\,d\zeta.$$

Recalling that

$$X(z) = \sum_{k=-\infty}^{m} C_{-k}(z - z_0)^{-k},$$

we can see that $X(z)(z - z_0)^m$ becomes analytic and can be expressed by Cauchy's integral theorem as

$$X(z)(z - z_0)^m = \frac{1}{2\pi j} \oint_{c_r} \frac{X(\zeta)(\zeta - z_0)^m}{\zeta - z} d\zeta,$$

where the contour integral of $X(z)$ along c_r is performed clockwise. Differentiating both sides of the equation above by z ($m - 1$) times leads to

$$\frac{d^{(m-1)}}{dz^{(m-1)}}[X(z)(z - z_0)^m] = \frac{(m-1)!}{2\pi j} \oint_{c_r} \frac{X(\zeta)(\zeta - z_0)^m}{(\zeta - z)^m} d\zeta.$$

Setting $z = z_0$ we get

$$\frac{d^{(m-1)}}{dz^{(m-1)}}[X(z)(z - z_0)^m]\bigg|_{z=z_0} = \frac{(m-1)!}{2\pi j} \oint_{c_r} X(\zeta) d\zeta = (m-1)!C_{-1}.$$

Therefore,

$$\text{Res}[X(z)|z = z_0] = C_{-1} = \frac{1}{(m-1)!} \frac{d^{(m-1)}}{dz^{(m-1)}}[X(z)(z - z_0)^m]\bigg|_{z=z_0}.$$

Example 3.3.1

Let us try to find the inverse z-transform of $S(z)$ obtained in Example 3.1.1 by the contour integral

$$S(z) = \frac{1}{1 - az^{-1}} = \frac{z}{z - a} \qquad |z| > |a|.$$

The inverse z-transform is computed as follows:

$$s(n) = \frac{1}{2\pi j} \oint_c S(z)z^{n-1} dz = \frac{1}{2\pi j} \oint_c \frac{z}{z - a} z^{n-1} dz.$$

If $n \geq 0$, the integrand of the contour integral has only one pole at $z = a$, as shown in Fig. 3.3.3(a). If we choose the integral path c as a circle with radius $R > |a|$ contained in the region of convergence, $s(n)$ can be obtained as

$$s(n) = \text{Res}\left[\frac{z}{z - a} z^{n-1}\bigg| z = a\right] = (z - a)\frac{z}{z - a} z^{n-1}\bigg|_{z=a}$$

$$= a \times a^{n-1} = a^n.$$

$$S(z) = \frac{z}{z-a} \quad |z| > |a|$$

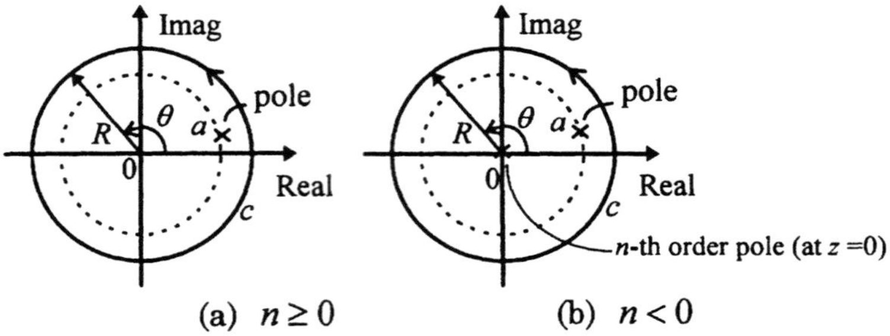

(a) $n \geq 0$ (b) $n < 0$

Figure 3.3.3   The pole locations of $S(z)$ when $n \geq 0$ (a) and $n < 0$ (b). In the case of $n \geq 0$, there is a pole at $z = a$ inside the loop c. In the case of $n < 0$, the order of the pole at $z = 0$ increases as the order of decreasing n.

The case where $n < 0$ is not as easy as $n \geq 0$, because an nth order pole of the integrand

$$\frac{z}{z-a}z^{n-1} = \frac{1}{z-a}z^{n},$$

is located at $z = 0$, and for increasing magnitude of n with the negative sign it will be more cumbersome to calculate the derivatives of the integrand. Let us try to get inverse z-transforms for $n = -1$ and $n = -2$. The integrand becomes

$$\frac{1}{z-a}z^{-1} \qquad \text{for } n = -1,$$

and

$$\frac{1}{z-a}z^{-2} \qquad \text{for } n = -2.$$

Since there are two poles inside c for $n = -1$ which are located at $z = a$ and $z = 0$, as shown in Fig. 3.3.3(b), $s(-1)$ is given by

$$s(-1) = \frac{1}{2\pi j} \oint_c \frac{1}{z-a}z^{-1} \, dz = \text{Res}\left[\frac{z}{z-a}z^{-2}\Big| z = a\right]$$

$$+ \text{Res}\left[\frac{z}{z-a}z^{-2}\Big| z = 0\right].$$

Executing residue calculations for the poles $z = a$ and $z = 0$, we get

$$\text{Res}\left[\frac{z}{z-a}z^{-2}\Big| z = a\right] = (z-a)\frac{z^{-1}}{z-a}\Bigg|_{z=a} = a^{-1}$$

and

$$\text{Res}\left[\frac{z}{z-a}z^{n-1}\bigg| z=0\right] = z\frac{z^{-1}}{z-a}\bigg|_{z=0} = -a^{-1}.$$

Therefore,

$$s(-1) = a^{-1} - a^{-1} = 0.$$

When $n = -2$, the integrand

$$\frac{z}{z-a}z^{n-1}\bigg|_{n=-2} = \frac{1}{(z-a)z^2}$$

and has two poles inside c, i.e., $z = a$ and $z = 0$, where the latter is a second-order pole. Therefore $s(-2)$ is obtained by summing the two residues corresponding to the two poles. Thus we have

$$s(-2) = \text{Res}\left[\frac{1}{(z-a)z^2}\bigg| z=a\right] + \text{Res}\left[\frac{1}{(z-a)z^2}\bigg| z=0\right].$$

The first term is obtained by the residue calculation as

$$\text{first term} = \frac{1}{a^2}.$$

But the second term requires more computational effort. We get

$$\text{second term} = \text{Res}\left[\frac{1}{(z-a)z^2}\bigg| z=0\right] = \frac{d}{dz}\left[z^2\frac{1}{(z-a)z^2}\right]\bigg|_{z=0}$$

$$= \frac{-1}{(z-a)^2}\bigg|_{z=0} = -\frac{1}{a^2}.$$

Therefore,

$$s(-2) = \frac{1}{a^2} - \frac{1}{a^2} = 0.$$

For negative integers less than -2, we must perform more tiresome work for the calculation of derivatives. We omit the calculation, but $s(n)$ for n less than -2 is zero as shown later. As a result, the inverse z-transform of $s(z)$ is

$$s(n) = \begin{cases} a^n & \text{for } n \geq 0 \\ 0 & \text{for } n < 0. \end{cases}$$

Comparing the sequence $s(n)$ obtained above with the sequence $s_1(n)$ found in Example 3.1.1, we see that $s(n)$ is identical to $s_1(n)$.

Example 3.3.2

Consider the inverse z-transform $S(z)$ identical to that obtained in Example 3.1.2,

$$S(z) = \frac{1}{1 - az^{-1}} = \frac{z}{z - a} \qquad |z| < |a|.$$

The inverse z-transform is

$$s(n) = \frac{1}{2\pi j} \oint_c \frac{z}{z - a} z^{n-1} dz$$

where c is a closed curve chosen as an integral path that is included in the region of convergence $|z| < |a|$ and encircles $z = 0$. In this case the integrand has no pole inside c for $n \geq 0$, so that

$$s(n) = 0 \qquad \text{for } n \geq 0.$$

For $n < 0$, residue calculations are as cumbersome as in the previous Example 3.3.1, so we use another way of calculating the residue. The residue theorem gives us another relationship,

$$\sum_i \text{Res}[X(z)|z = z_i] = -\sum_k \text{Res}[X(z)|z = z_k] - \text{Res}[X(z)|z = \infty],$$

where the z_i are poles inside c and the z_k are poles outside c of $X(z)$. The residue of $X(z)$ at infinity is obtained from

$$\text{Res}[X(z)|z = \infty] \equiv zX(z)|_{z=\infty}.$$

Using this relation we get

$$s(n) = \text{Res}\left[\frac{z^n}{z - a}\bigg| z = 0\right] = -\text{Res}\left[\frac{z^n}{z - a}\bigg| z = a\right] - \text{Res}\left[\frac{z^n}{z - a}\bigg| z = \infty\right]$$

$$= -a^n - 0 = -a^n \qquad \text{for } n < 0$$

because there is only a pole at $z = a$ outside c. The resulting $s(n)$ is the same sequence as that found in Example 3.1.2.

Example 3.3.3

Consider the sum of two z-transforms, $S_1(z)$ and $S_2(z)$,

$$S(z) = S_1(z) + S_2(z) \qquad \text{for } |a_1| < |z| < |a_2|,$$

where

$$S_1(z) = \frac{1}{1 - a_1 z^{-1}} = \frac{z}{z - a_1} \qquad \text{for } |z| > |a_1|$$

and

$$S_2(z) = \frac{1}{1 - a_2 z^{-1}} = \frac{z}{z - a_2} \qquad \text{for } |z| < |a_2|.$$

The inverse z-transforms of $S_1(z)$ and $S_2(z)$ are discussed in Examples 3.3.1 and 3.3.2. Now we will deal with the inverse z-transform resulting from the addition of the two z-transforms. Here, if $|a_1| < |a_2|$ holds, there is a region of convergence on the z-plane, because the intersection of $|z| > |a_1|$ and $|z| < |a_2|$ exists as an annular region as shown in Fig. 3.3.4. We choose a closed curve c as illustrated in Fig. 3.3.4 which encircles $z = 0$ as the path of the contour integral to find the inverse z-transform. For this integral path c we have an inverse z-transform,

$$s(n) = \frac{1}{2\pi j} \oint_c S(z) z^{n-1} \, dz = \frac{1}{2\pi j} \oint_c \frac{z}{z - a_1} z^{n-1} \, dz$$

$$+ \frac{1}{2\pi j} \oint_c \frac{z}{z - a_2} z^{n-1} \, dz$$

$$= \text{Res}[S(z) z^{n-1} | z = a_1],$$

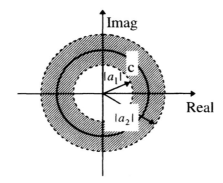

Figure 3.3.4 The path c of the contour integral in the region of convergence for inverse z-transform of $S(z)$ in Example 3.3.3.

because, for $n \geq 0$, the integrand of the first integral in the right-hand side has a single pole at $z = a_1$ inside c. But the integrand of the second integral

$$\frac{z}{z - a_2} z^{n-1}$$

has no poles inside c. Therefore,

$$s(n) = \text{Res}\left[\frac{z}{z - a_1} z^{n-1} \Big| z = a_1\right] = \text{Res}\left[\frac{z^n}{z - a_1} \Big| z = a_1\right] = a_1^n,$$

where $n \geq 0$.

When $n < 0$, the integrand of the first integral has two poles inside c, a pole at $z = a_1$ and an nth order pole at $z = 0$, and the integrand of the second integral has an nth order pole at $z = 0$ inside c. So we obtain the inverse z-transform of $S(z)$ as

$$s(n) = \frac{1}{2\pi j} \oint_c S(z) z^{n-1} \, dz$$

$$= \mathrm{Res}\left[(z - a_1) \frac{z^n}{z - a_1} \bigg| z = a_1 \right] + \mathrm{Res}\left[\frac{z^n}{z - a_1} \bigg| z = 0 \right]$$

$$+ \mathrm{Res}\left[\frac{z^n}{z - a_2} \bigg| z = 0 \right]$$

$$= (z - a_1) \frac{z^n}{(z - a_1)} \bigg|_{z=a_1} + \frac{1}{(|n| - 1)!} \frac{d^{|n|-1}}{dz^{|n|-1}} \left(z^{-n} \frac{z^n}{z - a_1} \right) \bigg|_{z=0}$$

$$+ \mathrm{Res}\left[\frac{z^n}{z - a_2} \bigg| z = 0 \right]$$

$$= a_1^n - a_1^{-|n|} + \mathrm{Res}\left[\frac{z^n}{z - a_2} \bigg| z = 0 \right]$$

$$= a_1^{-|n|} - a_1^{-|n|} - \mathrm{Res}\left[\frac{z^n}{z - a_2} \bigg| z = a_2 \right] - \mathrm{Res}\left[\frac{z^n}{z - a_2} \bigg| z = \infty \right]$$

$$= -a_2^n - 0 = -a_2^n.$$

Thus, we obtain

$$s(n) = \begin{cases} a_1^n & \text{for } n \geq 0 \\ -a_2^n & \text{for } n < 0. \end{cases}$$

3.3.2 Methods without Contour Integration

In the preceding section we discussed a mathematical method using contour integration for finding the inverse z-transforms. This is rather a formal and time-consuming procedure. We can get the inverse z-transforms without using mathematically formal procedures.

Consider the inverse z-transform of a single-pole system

$$H(z) = \frac{1}{1 - az^{-1}} \qquad |z| > |a|,$$

where a is a constant. We have already studied this type of z-transform in Example 3.3.1 and we know that the inverse z-transform $h(n)$ is

$$h(n) = \begin{cases} a^n & \text{for } n \geq 0 \\ 0 & \text{for } n < 0. \end{cases}$$

Thus, if we encounter a z-transform of a single-pole system we can obtain its inverse z-transform $h(n)$ recalling this example. This method is useful not only for the z-transform of a single pole but also for general rational functions of z. In a system expressed by a linear and constant coefficient difference equation, the input–output relationship in the z-plane is represented by a rational function which is given by the ratio of two polynomials of z^{-1}.

Consider a rational function

$$H(z) = \frac{b_0 + b_1 z^{-1} + \cdots + b_{M-1} z^{-(M-1)}}{a_0 + a_1 z^{-1} + \cdots + a_{N-1} z^{-(N-1)}} \qquad (N > M),$$

where the a_n and b_n are real constants. By factoring the denominator of $H(z)$ we get

$$H(z) = \frac{\displaystyle\sum_{j=0}^{M-1} b_j z^{-j}}{a_0 \displaystyle\prod_{i=1}^{N-1} (1 - z_i z^{-1})}.$$

Then $H(z)$ can be decomposed to a sum of partial fractions as

$$H(z) = \frac{\displaystyle\sum_{j=0}^{M-1} b_j z^{-j}}{a_0 \displaystyle\prod_{i=1}^{N-1} (1 - z_i z^{-1})} = \frac{1}{a_0} \sum_{i=1}^{N-1} \frac{A_i}{1 - z_i z^{-1}},$$

where the A_i are complex constants.

Assuming that $H(z)$ is the z-transform of a right-sided sequence, the region of convergence is the outer region of a circle with the radius R which is equal to the largest magnitude of the poles z_i, i.e.,

$$R = \max(|z_i|, i = 1, 2, \ldots, (N-1)).$$

Here the A_i are found by performing residue calculations for each term in the sum of partial fractions above, that is,

$$\mathrm{Res}[H(z)|z = z_i] = (z - z_i)H(z)|_{z=z_i} = z_i A_i.$$

If z_k and z_l are complex conjugate poles $(z_k = z_l^*)$, we get $A_k = A_l^*$. This is shown as follows. $H(z)$ satisfies the equality

$$H(z^*) = \frac{b_0 + b_1(z^{-1})^* + \cdots + b_{M-1}(z^{-(M-1)})^*}{a_0 + a_1(z^{-1})^* + \cdots + a_{N-1}(z^{-(N-1)})^*} = H^*(z).$$

$H(z)$ can be rewritten as

$$H(z) = \frac{R(z)}{(1 - z_k z^{-1})(1 - z_l z^{-1})},$$

where

$$R(z) = \frac{\sum_{j=0}^{M-1} b_j z^{-j}}{a_0 \prod_{i \neq k,l} (1 - z_i z^{-1})}.$$

Noting that

$$\{(1 - z_k z^{-1})(1 - z_l z^{-1})\}^* = (1 - z_k^* z^{-1*})(1 - z_l^* z^{-1*})$$
$$= (1 - z_l z^{-1*})(1 - z_k z^{-1*}),$$

we get

$$R^*(z) = H^*(z)\{(1 - z_k z^{-1})(1 - z_l z^{-1})\}^* = H(z^*)(1 - z_k z^{-1*})(1 - z_l z^{-1*})$$
$$= R(z^*).$$

A_k and A_l are written as

$$A_k = \frac{1}{z_k} \mathrm{Res}[H(z)|z = z_k] = \frac{1}{z_k}(z - z_k)H(z)\Big|_{z=z_k} = \frac{R(z_k)}{1 - z_l z_k^{-1}}$$

and

$$A_l = \frac{1}{z_l} \mathrm{Res}[H(z)|z = z_l] = \frac{1}{z_l}(z - z_l)H(z)\Big|_{z=z_l} = \frac{R(z_l)}{1 - z_k z_l^{-1}}.$$

Therefore, we get

$$A_k^* = \frac{R^*(z_k)}{1 - z_l^* z_k^{-1*}} = \frac{R(z_k^*)}{1 - z_k z_l^{-1}} = \frac{R(z_l)}{1 - z_k z_l^{-1}} = A_l.$$

Each term in the partial fraction expansion of $H(z)$ is a z-transform of a sequence of the form

$$h_i(n) = \begin{cases} A_i z_i^n & \text{for } n \geq 0 \\ 0 & \text{for } n < 0, \end{cases}$$

as shown in Example 3.3.1 or by recalling that

$$h(n) = \begin{cases} a^n & \text{for } n \geq 0 \\ 0 & \text{for } n < 0 \end{cases}$$

is the inverse z-transform of $H(z)$,

$$\frac{1}{1 - az^{-1}} \qquad \text{for } |z| > a.$$

Thus, the inverse z-transform $h(n)$ of $H(z)$,

$$H(z) = \frac{1}{a_0} \sum_{i=1}^{N-1} \frac{A_i}{1 - z_i z^{-1}}$$

is found by summing over the index i,

$$h(n) = \begin{cases} \dfrac{1}{a_0} \displaystyle\sum_{i=1}^{N-1} A_i z_i^n & \text{for } n \geq 0 \\ 0 & \text{for } n < 0. \end{cases}$$

Assuming that the time sequence $h(n)$ is real, the z_i consist of a pair of complex conjugate poles and real poles. If z_k and z_l are complex conjugate poles written as $z_k = z_l^*$, the sum of the corresponding two complex sequences $h_k(n)$ and $h_l(n)$ must be real, as follows:

$$h_k(n) + h_l(n) = A_k z_k^n + A_l z_l^n = A_k z_k^n + (A_k z_k^n)^* = 2\,\text{Re}(A_k z_k^n)$$

because $A_k^* = A_l$. When second- or higher-order poles can be found in the denominator of $H(z)$, we can use the residue calculation method as already discussed to get a partial fraction expansion. We next discuss some examples using partial fraction expansions.

Example 3.3.4

Consider a rational function $H(z)$ of the form

$$H(z) = \frac{1}{1 - 0.1z^{-1} - 0.2z^{-2}} \qquad |z| > |0.5|.$$

$H(z)$ can be decomposed into a partial fraction sum using residue calculation as

$$H(z) = \frac{1}{(1 - 0.5z^{-1})(1 + 0.4z^{-1})} = \frac{A_1}{1 - 0.5z^{-1}} + \frac{A_2}{1 + 0.4z^{-1}}$$

$$= \frac{5/9}{1 - 0.5z^{-1}} + \frac{4/9}{1 + 0.4z^{-1}},$$

where

$$A_1 = \text{Res}\left[\frac{1}{(1 - 0.5z^{-1})(1 + 0.4z^{-1})}\Big| z = 0.5\right] = \frac{1}{1 + (0.4/0.5)} = \frac{5}{9}$$

and

$$A_2 = \text{Res}\left[\frac{1}{(1 - 0.5z^{-1})(1 + 0.4z^{-1})}\Big| z = -0.4\right] = \frac{1}{1 + (5/4)} = \frac{4}{9}.$$

The two terms of the right-hand side are converted to two sequences $h_1(n)$ and $h_2(n)$, respectively, which are in turn added to get the final result,

$$h(n) = h_1(n) + h_2(n) = \begin{cases} \dfrac{5}{9} \times 0.5^n + \dfrac{4}{9} \times (-0.4)^n & \text{for } n \geq 0 \\ 0 & \text{for } n < 0. \end{cases}$$

Since $H(z)$ is a simple rational function, the partial expansion procedure can be carried out without using the residue calculation. We can get A_1 and A_2 by solving

$$H(z) = \frac{A_1}{1 - 0.5z^{-1}} + \frac{A_2}{1 + 0.4z^{-1}} = \frac{1}{(1 - 0.5z^{-1})(1 + 0.4z^{-1})}.$$

Example 3.3.5

Consider the following $H(z)$:

$$H(z) = \frac{1}{(1 - z_0 z^{-1})(1 - z_0^* z^{-1})} \qquad |z| > |z_0|,$$

where z_0^* denotes the complex conjugate of z_0. This $H(z)$ can be expressed by a partial expansion

$$H(z) = \frac{1}{(1 - z_0 z^{-1})(1 - z_0^* z^{-1})} = \frac{A}{1 - z_0 z^{-1}} + \frac{B}{1 - z_0^* z^{-1}}$$

$$= H_1(z) + H_2(z),$$

where A and B are obtained by solving

$$A(1 - z_0^* z^{-1}) + B(1 - z_0 z^*) = 1 \qquad \text{for any } z.$$

As a result we get

$$A = \frac{1}{1 - z_0^* z_0^{-1}} \qquad \text{and} \qquad B = \frac{1}{1 - z_0 (z_0^{-1})^*}.$$

Therefore, we can see that $A = B^*$. The inverse z-transforms $h_1(n)$ of

$$H_1(z) = \frac{A}{1 - z_0 z^{-1}}$$

and $h_2(n)$ of

$$H_2(z) = \frac{B}{1 - z_0^* z^{-1}}$$

are

$$h_1(n) = A z_0^n = \frac{z_0^n}{1 - z_0^* z_0^{-1}} = h_2^*(n)$$

and

$$h_2(n) = B(z_0^*)^n = \frac{(z_0^*)^n}{1 - z_0(z_0^{-1})^*} = h_1^*(n),$$

respectively. The sequences $h_1(n)$ and $h_2(n)$ can be derived from the result of Example 3.1.1. The inverse z-transform of $H(z)$ is

$$h(n) = h_1(n) + h_2(n) = h_1(n) + h_1^*(n) = h_2^*(n) + h_2(n)$$

$$= 2 \operatorname{Re} \left\{ \frac{z_0^n}{1 - z_0^* z_0^{-1}} \right\},$$

or equivalently

$$h(n) = 2 \operatorname{Re} \left\{ \frac{(z_0^*)^n}{1 - z_0(z_0^*)^{-1}} \right\} \qquad \text{for } n \geq 0.$$

Another way to get the inverse z-transform is by making use of a long division method, which is useful for $H(z)$ expressed as a rational function.

Example 3.3.6

Suppose that $H(z)$ is a rational function as follows:

$$H(z) = \frac{1}{1 - 0.1z^{-1} - 0.2z^{-2}} \qquad |z| > R_1,$$

where R_1 denotes the largest magnitude of the two poles. $H(z)$ is an identical function to the z-transform in Example 3.3.4. Long division is performed as

$$
\begin{array}{r}
1 + 0.1z^{-1} + 0.21z^{-2} \ldots\ldots \\[4pt]
1 - 0.1z^{-1} - 0.2z^{-2} \overline{\smash{\big)}\, 1 } \\
1 - 0.1z^{-1} - 0.2z^{-2} \\ \hline
0.1z^{-1} + 0.2z^{-2} \\
0.1z^{-1} - 0.01z^{-2} - 0.02z^{-3} \\ \hline
0.21z^{-2} + 0.02z^{-3} \\
0.21z^{-2} - 0.021z^{-3} - 0.042z^{-4} \\ \hline
0.041z^{-3} + 0.042z^{-4}.
\end{array}
$$

The inverse z-transform will be

$$H(z) = 1 + 0.1z^{-1} + 0.21z^{-2} + \cdots.$$

Therefore, we have the sequence

$$h(0) = 1, \quad h(1) = 0.1, \quad h(2) = 0.21, \ldots.$$

This is the same result as obtained in Example 3.3.4, as we expected.

If the region of convergence is the inner region of a circle $|z| = R_2$, then the inverse z-transform must be a left-sided sequence as discussed in Section 3.2.3, where R_2 is the smallest magnitude of the two poles of $H(z)$. Thus the long division for $H(z)$ has to be performed for increasing powers of z, the opposite of the case $|z| > R_1$ as follows:

$$
\begin{array}{r}
-5z^2 + 2.5z^3 - 26.25z^4 \ldots\ldots \\[4pt]
-0.2z^{-2} - 0.1z^{-1} + 1 \overline{\smash{\big)}\, 1 } \\
1 + 0.5z - 5z^2 \\ \hline
-0.5z + 5z^2 \\
-0.5z - 0.25z^2 + 2.5z^3 \\ \hline
5.25z^2 - 2.5z^3 \\
5.25z^2 + 2.625z^3 - 26.25z^4 \\ \hline
-5.125z^3 + 26.25z^4.
\end{array}
$$

We obtain the inverse transform as

$$h(-2) = -5, \quad h(-3) = 2.5, \quad h(-4) = -26.25, \ldots.$$

For the above examples it should be emphasized again that the difference in the regions of convergence is important in obtaining the correct result.

3.4 Properties of the z-Transform

The z-transform has many useful properties in analyzing discrete-time signals and systems. The z-transform properties are all derived from the z-transform definition. In the following sections, we exploit a notation indicating the relationship between a sequence and its z-transform as

$$s(n) \underset{IZT}{\overset{ZT}{\rightleftharpoons}} S(z).$$

The left-hand side sequence can be transformed to the right-hand side by z-transformation and vice versa. The arrow directed to the left means an inverse z-transform and the arrow to the right means the z-transform. ZT and IZT indicate z-transform and inverse z-transform, respectively.

3.4.1 Linearity

If the z-transforms of sequences $s_1(n)$ and $s_2(n)$ are $S_1(z)$ and $S_2(z)$, respectively, with a_1 and a_2 complex constants, then

$$a_1 s_1(n) + a_2 s_2(n) \underset{IZT}{\overset{ZT}{\rightleftharpoons}} a_1 S_1(z) + a_2 S_2(z)$$

holds where the region of convergence is the intersection of the convergence regions of $S_1(z)$ and $S_2(z)$.

The linearity is one of the most important properties and is particularly useful when a complicated rational function of z has to be decomposed into a sum of partial fractions in order to find the corresponding inverse z-transform. The proof of the linearity is found below. We can get

$$S_1(z) = \sum_{n=-\infty}^{\infty} s_1(n) z^{-n}$$

and

$$S_2(z) = \sum_{n=-\infty}^{\infty} s_2(n) z^{-n};$$

therefore,

$$a_1 S_1(z) + a_2 S_2(z) = a_1 \sum_{n=-\infty}^{\infty} s_1(n)z^{-n} + a_2 \sum_{n=-\infty}^{\infty} s_2(n)z^{-n}$$

$$= \sum_{n=-\infty}^{\infty} \{a_1 s_1(n) + a_2 s_2(n)\}z^{-n}.$$

3.4.2   Time Shift

This property says that if a sequence $s(n)$ is time-shifted by an integer constant m_0, the z-transform is as

$$s(n - m_0) \underset{IZT}{\overset{ZT}{\rightleftharpoons}} z^{-m_0} S(z),$$

where the region of convergence is equal to that of $S(z)$ except $z = 0$ (for $m_0 > 0$) or $z = \infty$ (for $m_0 < 0$). The proof is as follows. Since

$$s(n) \underset{IZT}{\overset{ZT}{\rightleftharpoons}} S(z)$$

the z-transform of the time-shifted sequence $s(n - m_0)$ is obtained by changing the time variable as $n - m_0 = k$,

$$\sum_{n=-\infty}^{\infty} s(n - m_0)z^{-n} = \sum_{k=-\infty}^{\infty} s(k)z^{-(k+m_0)} = z^{-m_0} \sum_{k=-\infty}^{\infty} s(k)z^{-k}$$

$$= z^{-m_0} S(z).$$

This property is convenient for obtaining the z-transform of a sum of sequences each of which is delayed by a different amount of time. For instance, consider $s(n)$ which has the z-transform $S(z)$. Let $s_1(n)$ be the sum of $s(n)$ and its delayed sequences, for example,

$$s_1(n) = s(n) + s(n - 1) + s(n - 2).$$

The z-transform of the summed sequence $s_1(n)$ is

$$\sum_{n=-\infty}^{\infty} s_1(n)z^{-n} = \sum_{n=-\infty}^{\infty} \{s(n) + s(n - 1) + s(n - 2)\}z^{-n}$$

$$= \sum_{n=-\infty}^{\infty} \{s(n)z^{-n} + s(n - 1)z^{-n} + s(n - 2)z^{-n}\}$$

$$= S(z) + z^{-1}S(z) + z^{-2}S(z)$$
$$= S(z)(1 + z^{-1} + z^{-2}),$$

where we have used the relations

$$\sum_{n=-\infty}^{+\infty} s(n-1)z^{-n} = z^{-1}S(z)$$

and

$$\sum_{n=-\infty}^{+\infty} s(n-2)z^{-n} = z^{-2}S(z).$$

In the derivation above, two z-transform properties are used, i.e., linearity and time shift. This example suggests that the z-transform of a linear and constant coefficient difference equation which will be introduced in Section 4.2 can be expressed in the z-plane by a rational function of z^{-1} multiplied by the z-transform of the sequence $s(n)$.

3.4.3   Exponential Weighting of a Sequence

This property shows that a sequence $s(n)$ weighted by e^{-n} has the z-transform as

$$e^{-n}s(n) \underset{IZT}{\overset{ZT}{\rightleftarrows}} S(ez).$$

That is, if the region of convergence of $s(n)$ is specified by $|z| > r$, then the resultant region of convergence for $e^{-n}s(n)$ is changed to

$$|z| > \frac{r}{e}.$$

The proof of this property is obtained as follows. The z-transform of $s(n)$ is written as

$$S(z) = \sum_{n=-\infty}^{\infty} s(n)z^{-n}$$

so that the z-transform of the weighted sequence $e^{-n}s(n)$ becomes

$$\sum_{n=-\infty}^{\infty} e^{-n}s(n)z^{-n} = \sum_{n=-\infty}^{\infty} s(n)(ez)^{-n} = S(ez).$$

Therefore the region of convergence $S(ez)$ is

$$|ez| > r, \text{ or } |z| > \frac{r}{e}.$$

3.4.4   Differentiation of the z-Transform

This property means that a sequence multiplied by time has the z-transform which is the differentiation of the original z-transform multiplied by $-z$. This is expressed by the following relation with respect to z:

$$ns(n) \underset{IZT}{\overset{ZT}{\rightleftharpoons}} -z\frac{dS(z)}{dz}.$$

The proof is shown as follows.
   By differentiating both sides of the relation

$$S(z) = \sum_{n=-\infty}^{\infty} s(n)z^{-n}$$

we get

$$\frac{dS(z)}{dz} = -\sum_{n=-\infty}^{\infty} ns(n)z^{-n-1} = -z^{-1}\sum_{n=-\infty}^{\infty} ns(n)z^{-n}$$

Therefore,

$$-z\frac{dS(z)}{dz} = \sum_{n=-\infty}^{\infty} ns(n)z^{-n}.$$

Thus we have

$$ns(n) \underset{IZT}{\overset{ZT}{\rightleftharpoons}} -z\frac{dS(z)}{dz}$$

The region of convergence is the same as the region for $S(z)$.

Example 3.4.1

Now consider a sequence

$$s(n) = \begin{cases} na^n & n \geq 0 \\ 0 & \text{otherwise.} \end{cases}$$

Since we know that the z-transform $S(z)$ of the sequence a^n for $n \geq 0$ is

$$S(z) = \frac{1}{1 - az^{-1}} \qquad |z| > |a|$$

we get

$$-z\frac{dS(z)}{dz} = -z\frac{(-1)(-a)(-1/z^2)}{(1-az^{-1})^2} = \frac{az^{-1}}{(1-az^{-1})^2}.$$

Therefore the z-transform of the sequence $s(n)$ is

$$\sum_{n=0}^{\infty} s(n)z^{-n} = \sum_{n=0}^{\infty} na^n z^{-n} = \frac{az^{-1}}{(1-az^{-1})^2}.$$

This differentiation property is convenient when we get the z-transform of a sequence of the form $ns(n)$.

3.4.5 Initial Value Theorem

This theorem shows that we can find the initial value of a sequence from a z-transform. Consider the z-transform $S(z)$ of a sequence $s(n)$. From the definition of the z-transform we can write

$$S(z) = \sum_{n=-\infty}^{\infty} s(n)z^{-n}.$$

Suppose that $s(n) = 0$ for $n < 0$ (this property is called causality of $s(n)$ as we will discuss in Section 4.3.2), then

$$\lim_{z\to\infty} S(z) = \lim_{z\to\infty} \sum_{n=0}^{\infty} s(n)z^{-n} = s(0)$$

because

$$\lim_{z\to\infty} z^{-n} = 0 \qquad \text{for } n > 1.$$

3.5 Convolution of Sequences

This property shows that a sequence $s_1(n)$ convolved with another sequence $s_2(n)$ has the z-transform as a product of the z-transforms $S_1(z)$ and $S_2(z)$,

$$s_1(n) * s_2(n) \underset{IZT}{\overset{ZT}{\rightleftharpoons}} S_1(z)S_2(z).$$

This relationship can be proved as follows. Let us take two z-transforms as

$$S_1(z) = \sum_{n=-\infty}^{\infty} s_1(n)z^{-n} \qquad |z| > R_1$$

and

$$S_2(z) = \sum_{n=-\infty}^{\infty} s_2(n)z^{-n} \qquad |z| > R_2.$$

From the definition of convolution discussed in Chapter 2, the z-transform of the convolved sequence $S_3(z)$ is obtained as

$$S_3(z) = \sum_{n=-\infty}^{\infty} \left\{ \sum_{m=-\infty}^{\infty} s_1(m)s_2(n - m) \right\} z^{-n}.$$

By changing the order of summation we get

$$S_3(z) = \sum_{m=-\infty}^{\infty} s_1(m) \sum_{n=-\infty}^{\infty} s_2(n - m)z^{-n}.$$

Again by changing the index as

$$n - m = k$$

the summation above becomes

$$S_3(z) = \sum_{m=-\infty}^{\infty} s_1(m) \sum_{k=-\infty}^{\infty} s_2(k)z^{-(m+k)}$$

$$= \sum_{m=-\infty}^{\infty} s_1(m)z^{-m} \sum_{k=-\infty}^{\infty} s_2(k)z^{-k}$$

$$= S_1(z) \cdot S_2(z).$$

The region of convergence of the product will be the overlap of the two regions of convergence for $S_1(z)$ and $S_2(z)$, because the product $S_1(z) \cdot S_2(z)$ exists only in the common region where both $S_1(z)$ and $S_2(z)$ exist.

3.6   Complex Convolution

In the previous section we saw that convolution of two sequences leads to a product of the two z-transforms. Reciprocally, does a product of two time sequences lead to a convolution of the two z-transforms? We have the result as

$$\sum_{n=-\infty}^{\infty} s_1(n) \cdot s_2(n)z^{-n} = \frac{1}{2\pi j} \oint_c S_1(v)S_2\left(\frac{z}{v}\right) v^{-1}\, dv.$$

Let us try to prove this relation. Let the left-hand side of the equation above be $W(z)$. Then

$$W(z) = \sum_{n=-\infty}^{\infty} s_1(n) \cdot s_2(n) z^{-n}$$

and

$$s_1(n) = \frac{1}{2\pi j} \oint_c S_1(v) v^{n-1} \, dv$$

This is the inverse z-transform definition found in Section 3.3, where c_1 is a closed curve within the region of convergence of $S_1(z)$ which is assumed to be of the form

$$R_{S1} < |z| < R_{L1}.$$

Substitution of the inverse z-transform of $s_1(n)$ into $W(z)$ leads to

$$W(z) = \sum_{n=-\infty}^{\infty} s_2(n) \left\{ \frac{1}{2\pi j} \oint_{c_1} S_1(v) v^{n-1} \, dv \right\} z^{-n}.$$

Changing the order of the summation and the contour integral, we get

$$W(z) = \frac{1}{2\pi j} \oint_{c_1} S_1(v) \left\{ \sum_{n=-\infty}^{\infty} s_2(n) \left(\frac{z}{v} \right)^{-n} \right\} \cdot \frac{1}{v} \, dv.$$

The term in braces,

$$\sum_{n=-\infty}^{\infty} s_2(n) \left(\frac{z}{v} \right)^{-n} = S_2 \left(\frac{z}{v} \right),$$

is the z-transform of $s_2(n)$ with the complex variable z/v. So we get

$$W(z) = \frac{1}{2\pi j} \oint_{c_2} S_1(v) \cdot S_2 \left(\frac{z}{v} \right) v^{-1} \, dv,$$

where c_2 is a closed curve in the region where both $S_1(z)$ and $S_2(z)$ exist.
   If the regions of convergence of $S_1(z)$ and $S_2(z)$ are specified, respectively, as

$$R_{S1} < |z| < R_{L1} \qquad \text{for } S_1(z)$$

and

$$R_{S2} < |z| < R_{L2} \qquad \text{for } S_2(z),$$

then the region of convergence of $S_2(z/v)$ is given by

$$R_{S2} < \left| \frac{z}{v} \right| < R_{L2},$$

or equivalently

$$R_{S1}R_{S2} < |z| < R_{L1}R_{L2}$$

where $R_{S1} < |v| < R_{L1}$.

The final result for $W(z)$ is not of a convolution form that we usually encounter in continuous-time systems. But if both the regions of convergence $S_1(z)$ and $S_2(z)$ include the unit circle, then c_2 can be chosen to be the unit circle so that the contour integral reduces to a line integral. That is, by setting

$$z = e^{j\omega} \qquad \text{and} \qquad v = e^{j\theta},$$

we get

$$
\begin{aligned}
W(e^{j\omega}) &= \frac{1}{2\pi j} \oint_{v=e^{j\theta}} S_1(v) S_2\left(\frac{z}{v}\right) v^{-1}\, dv \\
&= \frac{1}{2\pi j} \oint_{-\pi}^{\pi} S_1(e^{j\theta}) S_2(e^{j(\omega-\theta)}) e^{-j\theta} j e^{j\theta}\, d\theta \\
&= \frac{1}{2\pi} \int_{-\pi}^{\pi} S_1(e^{j\theta}) S_2(e^{j(\theta-\omega)})\, d\theta
\end{aligned}
$$

where the relationship

$$\frac{dv}{d\theta} = jv = je^{j\theta}$$

is used. $W(e^{j\omega})$ looks similar to the form of continuous-time convolution.

3.7   Parseval's Equation

Let us consider the energy of a real sequence $s(n)$. Energy Es of a real sequence is defined by the sum of squared samples, which is written as

$$Es = \sum_{n=-\infty}^{\infty} s^2(n).$$

By applying the complex convolution theorem to the sequence $s^2(n)$, we get

$$\sum_{n=-\infty}^{\infty} s^2(n) z^{-n} = \frac{1}{2\pi j} \oint_c S(v) S\left(\frac{z}{v}\right) v^{-1}\, dv.$$

Here if we set $z = 1$, we have

$$\sum_{n=-\infty}^{\infty} s^2(n) = \frac{1}{2\pi j} \oint_c S(v) S\left(\frac{1}{v}\right) v^{-1}\, dv = Es$$

where c is a closed curve in the region of convergence of $S(z)$. This relation is referred to as Parseval's equation.

If the unit circle is contained in the region of convergence, the contour integral can be replaced by a definite integral in the interval between $-\pi$ and $+\pi$,

$$\sum_{n=-\infty}^{\infty} s^2(n) = \frac{1}{2\pi} \int_{-\pi}^{\pi} S(e^{j\omega}) \cdot S(e^{-j\omega})\, d\omega$$

$$= \frac{1}{2\pi} \int_{-\pi}^{\pi} |S(e^{j\omega})|^2\, d\omega = \int_{-1/2}^{+1/2} |S(e^{j2\pi f})|^2\, df.$$

Here $\omega = 2\pi f$ and $|S(e^{j\omega})|^2$ or $|S(e^{j2\pi f})|^2$ is referred to as the power spectrum density of the sequence $s(n)$. The equation shows that the energy obtained by summing all squared samples is identical to the energy calculated by the integration of the power spectrum density over the entire (angular) frequency range (except for a constant $1/2\pi$).

Parseval's equation can be generalized to complex sequences. If $s(n)$ is a complex sequence, then the absolutely squared sum of $s(n)$ is expressed by the complex convolution as

$$E = \sum_{n=-\infty}^{\infty} |s(n)|^2 = \sum_{n=-\infty}^{\infty} s(n)s^*(n) = \frac{1}{2\pi j} \oint_c S(v)S^*\left(\frac{1}{v^*}\right) v^{-1}\, dv.$$

To show this relation we set

$$s_1(n) = s(n) \qquad \text{and} \qquad s_2(n) = s^*(n).$$

Using the complex convolution theorem, we have

$$\sum_{n=-\infty}^{\infty} s(n)s^*(n)z^{-n} = \frac{1}{2\pi j} \oint_c S(v)S^*\left(\frac{z^*}{v^*}\right) v^{-1}\, dv.$$

If we set $z = 1$, we obtain the equation to get E. If the unit circle $v = e^{j\omega}$ is included in the region of convergence of $S(z)$, then E reduces to

$$E = \frac{1}{2\pi} \int_{-\pi}^{\pi} |S(e^{j\omega})|^2\, d\omega = \sum_{n=-\infty}^{\infty} |s(n)|^2.$$

3.8   Summary

We showed the definition of the z-transform and demonstrated some examples of sequences and their z-transforms, and pointed out that a z-transform should be defined with the region of convergence on the z-plane. We also discussed the regions of convergence of four classes of sequences. We discussed the methods for finding the inverse z-transform by contour integral and other practical methods. We showed a list of typical time sequences and their corresponding z-transforms. Finally, we studied the properties of z-transforms, the relationship between the convolution sum of two sequences and the z-transform, and Parseval's equation.

4 Transfer Function and Frequency Response Function

4.0 Introduction

In this chapter we discuss the properties of linear time-invariant systems which are frequently encountered in analysis and design for practical applications. Linear time-invariant systems are characterized by the convolution sum. An input sequence $x(n)$ to a system whose impulse response sequence is $h(n)$ and the output sequence $y(n)$ are related by

$$y(n) = h(n) * x(n),$$

where $*$ indicates taking a convolution. From the discussion on the z-transform of a convolution sum in Chapter 3, we saw that the z-transforms of the sequences $y(n)$, $x(n)$, and $h(n)$ are connected by a product form,

$$Y(z) = H(z)X(z)$$

where

$$x(n) \overset{ZT}{\underset{IZT}{\rightleftarrows}} X(z)$$

$$y(n) \overset{ZT}{\underset{IZT}{\rightleftarrows}} Y(z)$$

and

$$h(n) \overset{ZT}{\underset{IZT}{\rightleftarrows}} H(z)$$

These relationships are depicted in Fig. 4.0.1.

The function $H(z)$ is called the transfer function or the system function of a linear and time-invariant system. Through this book we use the term *transfer function*. A transfer function evaluated on the unit circle on the z-plane is called a *frequency response function* of a system, which is

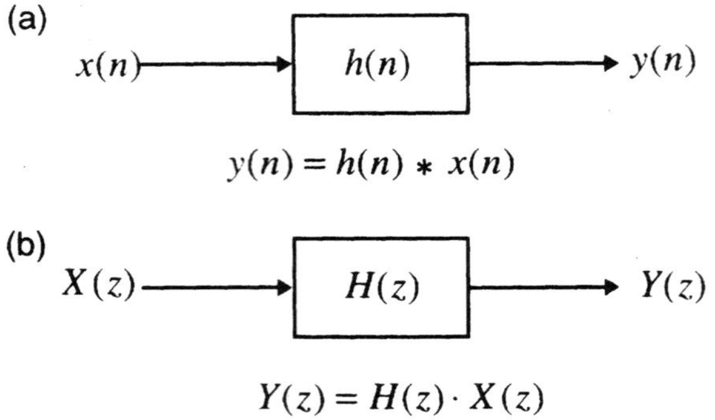

(a)

$$x(n) \longrightarrow \boxed{h(n)} \longrightarrow y(n)$$

$$y(n) = h(n) * x(n)$$

(b)

$$X(z) \longrightarrow \boxed{H(z)} \longrightarrow Y(z)$$

$$Y(z) = H(z) \cdot X(z)$$

Figure 4.0.1 A linear time-invariant system: (a) time-domain expression; (b) z-domain expression.

given by

$$H(e^{j\omega}) = H(z)\big|_{z=e^{j\omega}}.$$

$H(e^{j\omega})$ is obtained by replacing z in the transfer function $H(z)$ by $e^{j\omega}$.

Although $H(e^{j\omega})$ is a function of ω, the expression $H(e^{j\omega})$ is used to emphasize and clarify that it is a periodic function with a period 2π. It is seen by substituting ω in $H(e^{j\omega})$ by $\omega + 2\pi n$, provided that the unit circle is included in the region of convergence where $H(e^{j\omega})$ exists. By inspection we can see that $H(e^{j\omega})$ is nothing but the definition of the discrete-time Fourier transform of a sequence $h(n)$ as stated in Chapter 2. Throughout this section we denote the discrete-time Fourier transform of a sequence as $H(e^{j\omega})$ instead of $H_d(e^{j\omega})$.

4.1 Frequency Response Functions of Time-invariant Systems

4.1.1 Frequency Response

Consider a linear time-invariant system with the impulse response $h(n)$. The input is a complex exponential sequence with a single angular frequency ω_0 as illustrated in Fig. 4.1.1. Since the impulse response sequence of the system is $h(n)$, the output $y(n)$ from the system is obtained

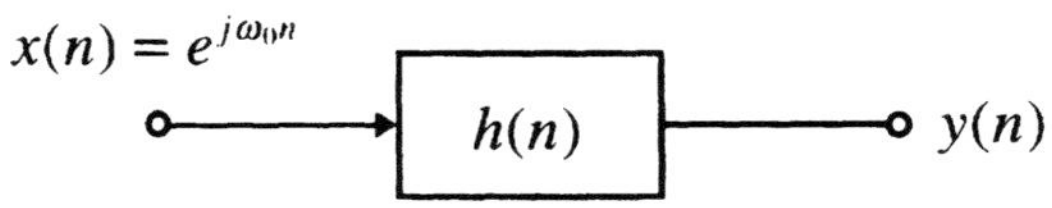

$$x(n) = e^{j\omega_0 n}$$
$$\circ \longrightarrow \boxed{h(n)} \longrightarrow \circ \ y(n)$$

Figure 4.1.1 A linear time-invariant system with the impulse response $h(n)$ and the input time sequence $e^{j\omega_0 n}$.

by the convolution sum of $h(n)$ and $e^{j\omega_0 n}$

$$y(n) = h(n) * e^{j\omega_0 n} = \sum_{k=-\infty}^{\infty} h(k)e^{j\omega_0(n-k)} = e^{j\omega_0 n} \sum_{k=-\infty}^{\infty} h(k)e^{-j\omega_0 k}$$

$$= e^{j\omega_0 n} \cdot H(e^{j\omega_0}).$$

This shows that the amplitude of a complex exponential sequence (or equivalently a sinusoidal sequence) with a single frequency ω_0 applied to the input of a linear time-invariant system is modified by the amount of $H(e^{j\omega_0})$. $H(e^{j\omega_0})$ is, in general, a complex quantity and is written in the form

$$H(e^{j\omega_0}) = |H(e^{j\omega_0})|e^{j\theta(e^{j\omega_0})}$$

where $|H(e^{j\omega_0})|$ and $\theta(e^{j\omega_0})$ are called the magnitude and phase responses, respectively. Accordingly the magnitude and phase of the input sequence are modified by $H(e^{j\omega_0})$ as follows:

	input $(e^{j\omega_0 n})$		output
magnitude	1	$\Rightarrow$	$|H(e^{j\omega_0})|$
phase (rad)	0	$\Rightarrow$	$\theta(e^{j\omega_0})$

The relations above hold not only for a single frequency complex input sequence but also for an input sequence composed of a weighted sum of complex sequences with L different frequencies ω_l. For the input sequence $x(n)$,

$$x(n) = \sum_{l=1}^{L} a_l e^{j\omega_l n},$$

the output sequence can be obtained by the convolution as

$$y(n) = \sum_{k=-\infty}^{\infty} h(k)x(n-k) = \sum_{k=-\infty}^{\infty} h(k) \sum_{l=1}^{L} a_l e^{j\omega_l(n-k)}$$

$$= \sum_{l=1}^{L} a_l e^{j\omega_l n} \sum_{k=-\infty}^{+\infty} h(k)e^{-j\omega_l k} = \sum_{l=1}^{L} a_l H(e^{j\omega_l}) \cdot e^{j\omega_l n}$$

where the a_l are weighting constants,

$$x(n-k) = \sum_{l=1}^{L} a_l e^{j\omega_l(n-k)}$$

and

$$H(e^{j\omega_l}) = \sum_{k=-\infty}^{+\infty} h(k)e^{-j\omega_l k}.$$

The output sequence $y(n)$, when the input sequence is the weighted sum of complex exponential sequences each of which has a different frequency, is seen to be a sum of all the complex sequences with each magnitude and phase modified by the frequency response $H(e^{j\omega_l})$.

In the same way as in the discussions above we can consider the system output when an input sequence is expressed as an integral form

$$x(n) = \int_R a(\omega)e^{j\omega n} d\omega,$$

provided that $x(n)$ exists, where R is an appropriate frequency range of ω. The output sequence $y(n)$ is obtained by the convolution sum of $x(n)$ and $h(n)$,

$$y(n) = \sum_{k=-\infty}^{\infty} h(k) \int_R a(\omega)e^{j\omega(n-k)} d\omega = \int_R a(\omega) \sum_{k=-\infty}^{\infty} h(k)e^{j\omega(n-k)} d\omega$$

$$= \int_R a(\omega) \left\{ \sum_{k=-\infty}^{+\infty} h(k)e^{-j\omega k} \right\} e^{j\omega n} d\omega = \int_R a(\omega)H(e^{j\omega})e^{j\omega n} d\omega.$$

From the result we can also see that each frequency component sequence

$$a(\omega)e^{j\omega n} d\omega$$

is modified by $H(e^{j\omega})$. The above discussions on the properties of transfer and frequency response functions are based on the principle of superposition of linear time-invariant systems.

4.1.2 Magnitude and Phase Responses

The magnitude of a frequency response function $H(e^{j\omega})$ is defined by the absolute value of $H(e^{j\omega})$. Since $H(e^{j\omega})$ is complex, $H(e^{j\omega})$ is expressed by a cartesian expression using the real and imaginary parts,

$$H(e^{j\omega}) = H_R(e^{j\omega}) + jH_I(e^{j\omega}),$$

where

$$H_R(e^{j\omega}) = \text{Re}\{H(e^{j\omega})\}$$

and

$$H_I(e^{j\omega}) = \text{Im}\{H(e^{j\omega})\}.$$

Therefore, the magnitude of $H(e^{j\omega})$ is

$$|H(e^{j\omega})| = \sqrt{H_R^2(e^{j\omega}) + H_I^2(e^{j\omega})},$$

where $H(e^{j\omega})$ is a vector on a complex plane from the origin to the point $H(e^{j\omega})$, so that the magnitude of $H(e^{j\omega})$ is the length of the vector.

We have another expression for $|H(e^{j\omega})|^2$ since the squared magnitude of a complex number is obtained as the number multiplied by its complex conjugate:

$$|H(e^{j\omega})|^2 = H(e^{j\omega}) \cdot H^*(e^{j\omega}).$$

In practical situations the magnitude is expressed by

$$10 \log_{10} |H(e^{j\omega})|^2,$$

which is called the log magnitude in dB.

On the other hand, the phase of $H(e^{j\omega})$ is defined by the angle of the vector $H(e^{j\omega})$ measured from the real positive axis to the vector anticlockwise. The phase response is

$$\theta(e^{j\omega}) = \tan^{-1} \frac{H_I(e^{j\omega})}{H_R(e^{j\omega})}.$$

Using the definition of $\theta(e^{j\omega})$ above, $H(e^{j\omega})$ can be represented by a polar expression as

$$H(e^{j\omega}) = |H(e^{j\omega})| \cdot e^{j\phi},$$

where

$$\phi = \theta(e^{j\omega}).$$

Taking the log of the expression above, we get

$$\ln\{H(e^{j\omega})\} = \ln\{|H(e^{j\omega})|\} + j\theta(e^{j\omega}).$$

Therefore, the phase response can also be represented by

$$\theta(e^{j\omega}) = \text{Im}[\ln\{H(e^{j\omega})\}].$$

In Fig. 4.1.2 we show the geometric expression of $H(e^{j\omega})$ by both cartesian and the polar coordinate systems. For convenience we sometimes use the symbol arg{·} to express phase response as

$$\arg\{H(e^{j\omega})\} = \theta(e^{j\omega}) = \tan^{-1} \frac{H_I(e^{j\omega})}{H_R(e^{j\omega})} = \text{Im}[\ln\{H(e^{j\omega})\}],$$

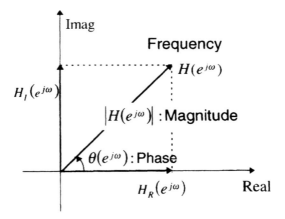

Figure 4.1.2   A geometric expression of a frequency response $H(e^{j\omega})$, where $H(e^{j\omega}) = H_R(e^{j\omega}) + jH_I(e^{j\omega})$ is a cartesian expression and $H(e^{j\omega}) = |H(e^{j\omega})|e^{d\theta(e^{j\omega})}$ is a polar expression.

where $\theta(e^{j\omega})$ is actually calculated making use of the arctan or complex log function but the functions can generate an angle in the interval $-\pi < \theta \le \pi$ in radians or $-180 < \theta \le 180$ in degrees.

The angle value within the interval is referred to as the 'principal value' of the phase functions. Since both arctan and complex log functions are multivalued functions, the exact values of these functions are

$$\tan^{-1}(*) = \theta \pm 2m\pi$$

where θ is a principal value and m is an arbitrary integer. This means that a phase value obtained for a particular ω has ambiguity by integer multiples of 2π.

When we consider periodic waveforms, principal values of the phase are useful, because we cannot distinguish between a periodic waveform and its waveform phase-shifted by $\pm 2m\pi$. But if we need a continuous phase function, the phase response composed of principal values must be changed to a continuous function by finding appropriate values of m for all ω. A phase sequence consists of principal values is called a 'wrapped' phase, while a continuous phase sequence generated from the wrapped phase is called an 'unwrapped' phase. The procedure for obtaining an unwrapped phase sequence is referred to as 'phase-unwrapping'.

The basic idea of phase unwrapping is as follows: When a phase change larger than $\pm\pi$ occurs between adjacent frequency points, we suppose that a $\pm 2m\pi$ jump arises and find an appropriate value of m by changing m until the difference of phase values of the two frequency points becomes less than $\pm\pi$. This approach is not very precise and a more sophisticated method is usually used.

In the following two examples, we examine the log magnitude in dB and the phase behavior for systems with a single pole and an all-zero frequency response.

Example 4.1.1

Consider a single-pole frequency response as

$$H(e^{j\omega}) = \frac{1}{1 - ae^{j\omega}}$$

$$= \frac{1 - a\cos\omega}{(1 - a\cos\omega)^2 + a^2\sin^2\omega} - j\frac{a\sin\omega}{(1 - a\cos\omega)^2 + a^2\sin^2\omega},$$

where a is a real positive number less than unity. This is an example of a frequency response function for a single-resonance system. The magnitude response of $H(e^{j\omega})$ is

$$|H(e^{j\omega})| = \frac{1}{|1 - ae^{-j\omega}|} = \frac{1}{\sqrt{(1 - ae^{-j\omega})(1 - ae^{-j\omega})}}$$

$$= \frac{1}{\sqrt{1 - ae^{j\omega} - ae^{-j\omega} + a^2}} = \frac{1}{\sqrt{1 - a(e^{-j\omega} + e^{-j\omega}) + a^2}}$$

$$= \frac{1}{\sqrt{1 - 2a\cos\omega + a^2}}.$$

The phase response of $H(e^{j\omega})$ is

$$\theta(e^{j\omega}) = \tan^{-1}\frac{-a\sin\omega}{1 - a\cos\omega}.$$

In Fig. 4.1.3 the magnitude response of $H(e^{j\omega})$ expressed in dB and the phase response $\theta(e^{j\omega})$ expressed in radians are illustrated together with the pole–zero location on the z-plane. From Fig. 4.1.3 we can see that the phase is a continuous function of ω and is included in the principal value range $-\pi \le \theta < \pi$.

Example 4.1.2

Consider a frequency response function expressed by an all-zero model,

$$H(e^{j\omega}) = (1 - a_1e^{-j\omega})(1 - a_1^*e^{-j\omega})(1 - a_2e^{-j\omega})(1 - a_2^*e^{-j\omega}),$$

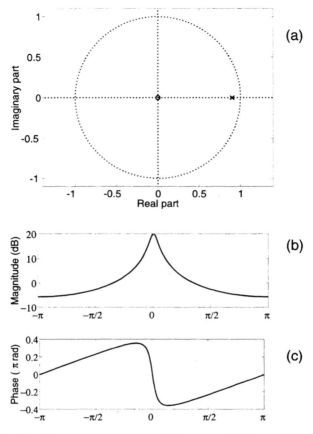

Figure 4.1.3   The frequency response of the system in Example 4.1.1: (a) pole and zero locations; (b) magnitude response; (c) phase response.

where a_k and a_k^* make a complex conjugate pair and $|a_k| = |a_k^*| > 1$. The logarithm of magnitude of $H(e^{j\omega})$ is

$$\ln |H(e^{j\omega})| = \ln |(1 - a_1 e^{-j\omega})(1 - a_1^* e^{-j\omega})(1 - a_2 e^{-j\omega})(1 - a_2^* e^{-j\omega})|$$

$$= \sum_{k=1}^{2} \ln |(1 - a_k e^{-j\omega})||(1 - a_k^* e^{-j\omega})|$$

and the phase response is

$$\theta(e^{j\omega}) = \text{Im}[\ln\{H(e^{j\omega})\}] = \sum_{k=1}^{2} \{\text{Im}(1 - a_k e^{-j\omega}) + \text{Im}(1 - a_k^* e^{-j\omega})\}.$$

    The magnitude and phase responses of $H(e^{j\omega})$ are depicted in Fig. 4.1.4 with the illustration of the pole–zero locations. In Fig. 4.1.4 we can see that the phase jump by $+2\pi$ occurs at the four frequency points. The phase response obtained through the unwrapping process is also depicted in Fig. 4.1.4, in which it is clear that the unwrapped phase response expands up to 8π radians within the frequency band $(-\pi \le \omega \le \pi)$.

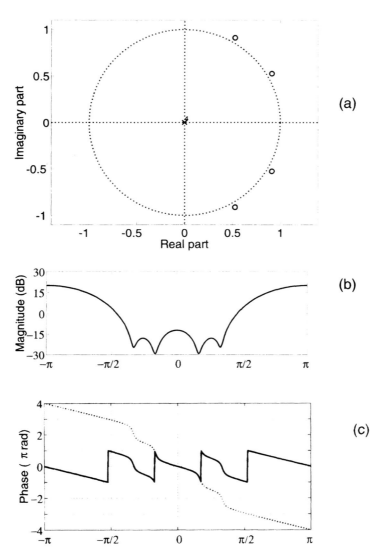

Figure 4.1.4   The frequency response of the system $H(e^{j\omega})$ in Example 4.1.2: (a) pole and zero locations; (b) magnitude response; (c) the wrapped phase (solid line) and unwrapped phase curves (dotted line).

4.2   Systems Represented by a Linear Constant Coefficient Difference Equation

4.2.1   Transfer Function Expressed by Rational Functions

In this section we consider the properties of a system characterized by a linear constant coefficient difference equation. The input and output relationship of the system is described by a difference equation such as

$$\sum_{k=0}^{K-1} a_k y(n-k) = \sum_{l=0}^{L-1} b_l x(n-l),$$

where $y(n)$ and $x(n)$ are the input and output sequences, respectively. The coefficients a_k and b_l are real constants characterizing the system. Without loss of generality we can set $a_0 = 1$ so that the difference equation above may be written as

$$y(n) = -\sum_{k=1}^{K-1} a_k y(n-k) + \sum_{l=0}^{L-1} b_l x(n-l),$$

where

$$\sum_{k=0}^{K-1} a_k y(n-k) = a_0 y(n) + \sum_{k=1}^{K-1} a_k y(n-k) = y(n) + \sum_{k=1}^{K-1} a_k y(n-k).$$

From the equation we see the present output sample value $y(n)$ is computed by the weighted sum of the L present and past input values $x(n)$ minus the weighted sum of the $(K-1)$ past output values $y(n)$. Taking z-transforms of both sides of the difference equation, we obtain the z-domain expression of the system as

$$Y(z) = -\sum_{k=1}^{K-1} a_k Y(z) z^{-k} + \sum_{l=0}^{L-1} b_l X(z) z^{-l}$$

$$= -Y(z) \sum_{k=1}^{K-1} a_k z^{-k} + X(z) \sum_{l=0}^{L-1} b_l z^{-l},$$

where in the derivation of this equation we use the time-shift property of z-transform.

As shown in Section 3.3.2, the transfer function $H(z)$ of the system is obtained by the ratio of $Y(z)$ to $X(z)$ as

$$H(z) = \frac{Y(z)}{X(z)} = \frac{\displaystyle\sum_{l=0}^{L-1} b_l z^{-l}}{\displaystyle\sum_{k=0}^{K-1} a_k z^{-k}}.$$

This shows that the transfer function of a system whose input–output relationship is represented by a linear constant coefficient difference equation is a ratio of polynomials of z^{-1}, i.e., a rational function of z^{-1}.

In general an Nth order polynomial can be factorized to a product of the N first-order factors. Therefore, we have another expression for the transfer function:

$$H(z) = \frac{b_0 \displaystyle\prod_{l=1}^{L-1}(1 - d_l z^{-1})}{\displaystyle\prod_{k=1}^{K-1}(1 - c_k z^{-1})}$$

provided that $a_0 = 1$. Here if several factors are of an identical form, they are counted separately. Since the c_k are zeros of the denominator, which are equivalent to the poles of $H(z)$, and the d_l are zeros of the numerator or $H(z)$, the transfer function expressed in this form is referred to as a pole–zero representation (or model) of the transfer function.

If the numerator reduces to a constant, $H(z)$ consists of only poles. In this case $H(z)$ is called an *all-pole* representation (or model). Conversely, if the denominator is a constant, then $H(z)$ consists of zeros and is called an *all-zero* representation (or model). The transfer function representations are used as system models for identification or design of acoustic systems in practical situations.

4.2.2 Pole–Zero Plot of Transfer Functions

It is convenient to express the transfer function of a system by plotting positions of poles and zeros of the transfer function. This plot is called a pole–zero plot. A transfer function of a system which is characterized by a linear and constant coefficient difference equation is expressed by a rational function or a pole–zero form as shown in Section 4.2.1.

Figure 4.2.1 shows an example of the pole–zero plot of the transfer function

$$H(z) = \frac{(1 - d_1 z^{-1})(1 - d_2 z^{-1})(1 - d_2^* z^{-1})}{(1 - c_1 z^{-1})(1 - c_2 z^{-1})(1 - c_2^* z^{-1})}$$

$$= \frac{(z - d_1)(z - d_2)(z - d_2^*)}{(z - c_1)(z - c_2)(z - c_2^*)} .$$

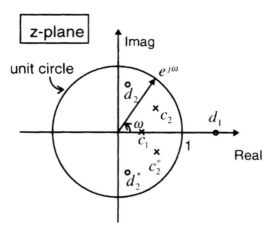

Figure 4.2.1 Pole–zero plot of a transfer function with 3 zeros and 3 poles.

This transfer function has one real (c_1) and one complex conjugate pair (c_2, c_2^*) of poles and one real (d_1) and one complex conjugate pair (d_2, d_2^*) of zeros. It is useful to show the unit circle on the same plane. This is because the frequency response function of the system

$$H(e^{j\omega}) = H(z)|_{z=e^{j\omega}}$$

exists if the unit circle is included in the region of convergence of $H(z)$. Using the pole–zero plot given in Fig. 4.2.1, consider the system properties such as the magnitude and phase responses and stability based on a geometrical interpretation.

4.2.3 Geometrical Interpretation of Magnitude Response

The magnitude response $H(e^{j\omega})$ of the example given in Fig. 4.2.1 is expressed by the definition

$$|H(e^{j\omega})| = \left| \frac{(1 - d_1 e^{-j\omega}) \cdot (1 - d_2 e^{-j\omega}) \cdot (1 - d_2^* e^{-j\omega})}{(1 - c_1 e^{-j\omega}) \cdot (1 - c_2 e^{-j\omega}) \cdot (1 - c_2^* e^{-j\omega})} \right|$$

$$= \frac{|(1 - d_1 e^{-j\omega})| \cdot |(1 - d_2 e^{-j\omega})| \cdot |(1 - d_2^* e^{-j\omega})|}{|(1 - c_1 e^{-j\omega})| \cdot |(1 - c_2 e^{-j\omega})| \cdot |(1 - c_2^* e^{-j\omega})|},$$

where $|H(e^{j\omega})|$ is composed of identical elementary terms of the form

$$H_a(e^{j\omega}) = (1 - ae^{-j\omega}).$$

Therefore, we start by investigating $H_a(e^{j\omega})$ where a is complex.
The magnitude response of $H(e^{j\omega})$ is

$$|H_a(e^{j\omega})| = |1 - ae^{-j\omega}| = |e^{-j\omega}(e^{j\omega} - a)| = |e^{j\omega} - a|.$$

The pole–zero plot is illustrated in Fig. 4.2.2. Accordingly, $|H_a(e^{j\omega})|$ means the distance or length on the z-plane from a particular point $e^{j\omega}$ on the unit circle corresponding to angular frequency ω to the point a as indicated in Fig. 4.2.2. In Fig. 4.2.2 the two cases $|a| < 1$ and $|a| > 1$ are illustrated. The distance $|e^{j\omega} - a|$ changes as ω goes on along the unit circle from $\omega = 0$ to 2π. This distance becomes smallest when $e^{j\omega}$ comes to the point closest to the point a. If a is a pole, the magnitude is the reciprocal of $|H_a(e^{j\omega})|$, so the inverse situation occurs. That is, when $e^{j\omega}$ reaches to the closest point on the unit circle to the point a (pole), the magnitude of $|H_a(e^{j\omega})|^{-1}$ has the largest distance.

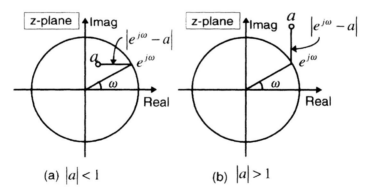

(a) $|a| < 1$ (b) $|a| > 1$

Figure 4.2.2 Illustrations of the magnitude response $|e^{j\omega} - a|$ of a single-zero system.

Since $H(z)$ illustrated in Fig. 4.2.1 has three zeros and three poles, the magnitude response of $H(e^{j\omega})$ is calculated by taking the product of the lengths from all the zeros to a particular point $e^{j\omega}$ on the unit circle corresponding to ω and dividing by lengths from all the poles to the same point. Thus, as ω goes anticlockwise along the unit circle starting at $z = 1$ ($\omega = 0$) the magnitude of $H(e^{j\omega})$ takes the locally smallest value each time $e^{j\omega}$ approaches the zero location, and takes the locally largest value when $e^{j\omega}$ gets close to the pole location. This accounts for peaks and dips appearing in the magnitude response of $H(e^{j\omega})$. The log

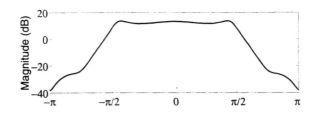

Figure 4.2.3   An example of the log magnitude response period.

magnitude response of $H(e^{j\omega})$ is illustrated in Fig. 4.2.3, where $c_1 = 0.8$, $c_2 = 0.9e^{j\pi/2}$, $d_1 = -0.9$, and $d_2 = 0.8e^{j3\pi/4}$.

4.2.4   Geometrical Interpretation of Phase Response

Next consider phase responses. The phase response of the system $H(e^{j\omega})$ is obtained by taking the imaginary part of the complex log of $H(e^{j\omega})$ as

$$\theta(e^{j\omega}) = \text{Im}\{\ln H(e^{j\omega})\}$$
$$= \text{Im}\{\ln(1 - d_1 e^{-j\omega}) + \ln(1 - d_2 e^{-j\omega}) + \ln(1 - d_2^* e^{-j\omega})$$
$$- \ln(1 - c_1 e^{-j\omega}) - \ln(1 - c_2 e^{-j\omega}) - \ln(1 - c_2^* e^{-j\omega})\},$$

where ln indicates the complex natural logarithm. This relation means that the total phase of $H(e^{j\omega})$ at a particular frequency ω is obtained by adding phase values corresponding to the individual zeros and subtracting phase values corresponding to the individual poles. It is helpful for us to study phase responses of single-zero and single-pole systems.

Let us consider the phase responses for single-zero and single-pole transfer functions. Figure 4.2.4 illustrates the three pole–zero plots corresponding to transfer functions consisting of (a) a single zero inside the unit circle, (b) a single zero outside the unit circle, and (c) a single pole inside the unit circle. The three transfer functions are given as

$$H_a(z) = 1 - z_0 z^{-1} \qquad (|z_0| < 1),$$
$$H_b(z) = 1 - z_0 z^{-1} \qquad (|z_0| > 1),$$
$$H_c(z) = \frac{1}{1 - z_0 z^{-1}} \qquad (|z_0| < 1).$$

The corresponding phase response of $H_a(z)$, $H_b(z)$, and $H_c(z)$ are illustrated in Fig. 4.2.4. The phase angle of $H_a(z)$ in Fig. 4.2.4(a) is

$$\phi_a(\omega) = \theta - \omega,$$

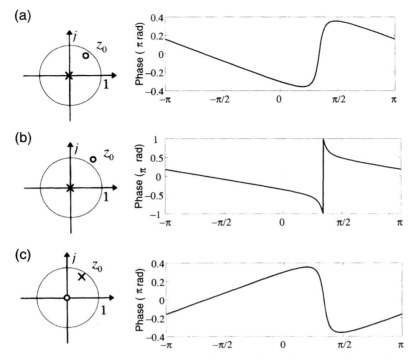

Figure 4.2.4 Phase curves corresponding to the pole–zero plots where the phase curves all vary rapidly at ω close to ω_0: (a) $1 - z_0 z^{-1}$, $(|z_0| < 1)$; (b) $1 - z_0 z^{-1}$, $(|z_0| > 1)$; (c) $1/1 - z_0 z^{-1}$, $(|z_0| < 1)$, $z_0 = r e^{j\omega_0}$.

where angles θ and ω are measured anticlockwise from the real positive axis, as shown in Fig. 4.2.5(a). This angle $\phi_a(\omega)$ is obtained as follows.

The frequency response function $H_a(e^{j\omega})$ is

$$H_a(e^{j\omega}) = (1 - z_0 e^{-j\omega}) = e^{-j\omega}(e^{j\omega} - z_0).$$

The phase angle of $H_a(e^{j\omega})$ is

$$\phi_a(\omega) = \arg\{H_a(e^{j\omega})\} = \operatorname{Im}\{\ln(e^{-j\omega}) + \ln(e^{j\omega} - z_0)\} = -\omega + \theta,$$

where, as shown in Fig. 4.2.5(a), θ is the angle of the vector $(e^{j\omega} - z_0)$ and $-\omega$ is the angle of the vector $e^{-j\omega}$. In a similar manner we can get the phase values for ϕ_b and ϕ_c as

$$\phi_b(\omega) = \theta - \omega$$

and

$$\phi_c(\omega) = \omega - \theta,$$

as shown in Fig. 4.2.5(b) and (c).

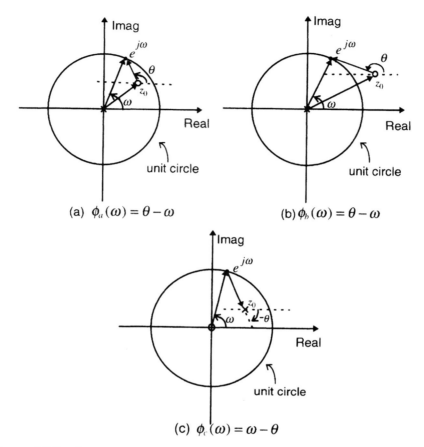

(a) $\phi_a(\omega) = \theta - \omega$ (b) $\phi_b(\omega) = \theta - \omega$

(c) $\phi_c(\omega) = \omega - \theta$

Figure 4.2.5   Geometric expressions of the phase functions: (a) $H_a(z) = 1 - z_0z^{-1}$, $(|z_0| < 1)$;   (b) $H_b(z) = 1 - z_0z^{-1}$, $(|z_0| > 1)$;   (c) $H_c(z) = 1/1 - z_0z^{-1}$, $(|z_0| < 1)$.

The phase response functions for the three cases illustrated in Fig. 4.2.4 are not symmetric with respect to $\omega = 0$ in the frequency range $-\pi \leq \omega < \pi$. This is because each one of the systems $H_a(\omega)$, $H_b(\omega)$, and $H_c(\omega)$ has no conjugate pair of the pole or zero. When the pole or zero is real, the phase function is an odd function. A system with a real impulse response or a real sequence always has complex conjugate pairs and has an odd phase function.

4.3   Stability and Causality of Systems

It is important for system identification and design to make decisions on the realizability of systems from the properties of z-transforms. This involves whether or not a system that has a transfer function expressed by a z-transform can be realized with the expected performance. There are two properties to be considered for system realizability: stability and causality of systems.

4.3.1 Stability of Systems

Consider the stability of a system with the transfer function $H(z)$. Let the impulse response sequence of the system $H(z)$ be $h(n)$. If $h(n)$ is absolutely summable,

$$\sum_{n=-\infty}^{\infty} |h(n)| < \infty,$$

then the system is called *stable*. As discussed in Section 4.1, the frequency response function $H(e^{j\omega})$ of a system with the transfer function $H(z)$ is written as

$$H(e^{j\omega}) = H(z)\big|_{z=e^{j\omega}}.$$

This means that in order for the system $H(z)$ to be stable the region of convergence of $H(z)$ includes the unit circle. When $h(n)$ is absolutely summable, we have

$$|H(e^{j\omega})| = \left|\sum h(n)e^{-j\omega n}\right| \le \sum |h(n)| \cdot |e^{-j\omega n}| < \sum |h(n)| < \infty.$$

Therefore $H(z)$ exists on the unit circle.

Example 4.3.1

Examine a system with the transfer function

$$H(z) = \frac{1}{1 - az^{-1}} \qquad |z| > |a|,$$

where a is a constant. If the sequence $h(n)$ corresponding to $H(z)$ is assumed to be a right-sided sequence,

$$h(n) = \begin{cases} a^n & n \ge 0 \\ 0 & \text{otherwise.} \end{cases}$$

If $|a| < 1$, the absolute sum of $h(n)$ exists as

$$\sum_{n=-\infty}^{\infty} |h(n)| = \sum_{n=0}^{\infty} |a^n| = \frac{1}{1 - |a|},$$

and the system is stable. On the other hand, if $|a| > 1$, the absolute sum of $h(n)$ does not exist and the system is not stable. If we relax $h(n)$ to be a left-sided sequence, we can get a stable sequence as

$$h(n) = \begin{cases} 0 & n \geq 0 \\ -a^n & n < 0. \end{cases}$$

4.3.2   Causality of Systems

Next we consider the causality of systems. A system is called *causal* when the output sequence begins after the initial input sample comes into a system. Let $h(n)$ and $H(z)$ be the impulse response sequence of a system and the corresponding transfer function, respectively. When the input sequence is $\delta(n - n_0)$ and the impulse response is

$$h(n - n_0) = \begin{cases} h(n - n_0) & \text{for } n \geq n_0 \\ 0 & \text{for } n < n_0, \end{cases}$$

the system is called causal, where n_0 is any integer.

Whether a system (or a sequence) is causal can be determined from $H(z)$. From the following properties of the z-transform of the sequence $h(n)$ $(n \geq 0)$:

$$\lim_{z \to \infty} H(z) = \lim_{z \to \infty} \{h(0) + h(1)z^{-1} + h(2)z^{-2} + \cdots\}$$
$$= h(0),$$

if $h(0)$ is a finite value, the system is causal.

Example 4.3.2

Consider $H(z)$ of the form

$$H(z) = z + \frac{1}{1 - az^{-1}} \qquad |z| > |a|,$$

where a is a constant. In this case $H(z)$ diverges with $z \to \infty$ because of the first term z of the right-hand side of the equation. The fact that $h(n)$ corresponding to the $H(z)$ is not causal is also confirmed by expanding the $H(z)$ in a power series. $H(z)$ is expanded as

$$H(z) = z + 1 + az^{-1} + a^2z^{-2} + \cdots$$

and the impulse response $h(n)$ is

$$h(n) = \begin{cases} 0 & n < -1 \\ 1 & n = -1 \\ a^n & n \geq 0. \end{cases}$$

We can see that an output sequence starts at $n = -1$ for the input impulse applied to the system at $n = 0$. Therefore the system is noncausal.

Example 4.3.3

Consider $H(z)$ of the form

$$H(z) = \frac{1}{1 - az^{-1}} \qquad |z| < |a|,$$

where a is a constant. Since $|az^{-1}| > 1$, $H(z)$ cannot be expanded in a power series in z^{-1}. If we change the form of $H(z)$ as

$$H(z) = \frac{1}{1 - az^{-1}} = \frac{z}{z - a} = \frac{a^{-1}z}{za^{-1} - 1} = \frac{-a^{-1}z}{1 - a^{-1}z},$$

then $H(z)$ can be expressed by a power series in z as

$$H(z) = -a^{-1}z(1 + a^{-1}z + a^{-2}z^2 + \cdots) = -a^{-1}z - a^{-2}z^2 - a^{-3}z^3 - \cdots.$$

The inverse z-transform $h(n)$ is

$$h(n) = \begin{cases} -a^n & \text{for } n \le -1 \\ 0 & \text{for } n \ge 0. \end{cases}$$

So $h(n)$ is not causal. If $a > 1$, then

$$\sum_{n=-\infty}^{\infty} |h(n)| < \infty$$

holds and thus the Fourier transform exists. Therefore $h(n)$ is a stable but noncausal sequence.

Since causality and stability are independent conditions of systems, there could be systems which are stable but noncausal, unstable but causal, and unstable and noncausal, none of which is realizable. Examples of the sequences with these conditions are illustrated in Fig. 4.3.1.

4.4 Relationships Between Real and Imaginary Parts of Frequency Response Functions

4.4.1 Real Sequences

Consider the Fourier transform of a real sequence $s(n)$,

$$S(e^{j\omega}) = \sum_{n=-\infty}^{\infty} s(n)e^{-j\omega n}$$

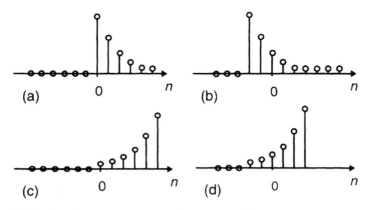

Figure 4.3.1   Impulse response sequences: (a) stable and causal; (b) stable but noncausal; (c) causal but unstable; (d) noncausal and unstable.

$$= \sum_{n=-\infty}^{\infty} s(n)\cos(\omega n) - j \sum_{n=-\infty}^{\infty} s(n)\sin(\omega n)$$

$$= S_R(e^{j\omega}) - jS_I(e^{j\omega}),$$

where $S_R(e^{j\omega})$ is a real even function of ω and $S_I(e^{j\omega})$ is a real odd function of ω, because $\cos(\omega n)$ and $\sin(\omega n)$ are even and odd functions of ω, respectively. This is correct only when $s(n)$ is a real sequence. Therefore, it should be noted that the real part of a transfer function for a real impulse response sequence must be an even function of ω and the imaginary part must be an odd function.

4.4.2   Real and Causal Sequences

Consider more restricted real sequences, which are sequences such as

$$s(n) = \begin{cases} s(n) & \text{for } n \geq 0 \\ 0 & \text{for } n < 0. \end{cases}$$

Such sequences are causal and divided into an even and an odd sequence $s_e(n)$ and $s_o(n)$, respectively,

$$s(n) = s_e(n) + s_o(n)$$

where

$$s_e(n) = \tfrac{1}{2}\{s(n) + s(-n)\} \qquad \text{for all } n$$

and

$$s_o(n) = \tfrac{1}{2}\{s(n) - s(-n)\} \qquad \text{for all } n.$$

Examples of the above sequences are shown in Fig. 4.4.1. From the equations above we can see that it is possible to derive the odd sequence component $s_o(n)$ from the even sequence component $s_e(n)$ and vice versa. Indeed, using the above equations we can get the following relationships:

$$s_e(n) = \begin{cases} s_o(n) & \text{for } n > 0 \\ -s_o(n) & \text{for } n < 0 \\ s(0) & \text{for } n = 0 \end{cases}$$

(a) $s(n)$

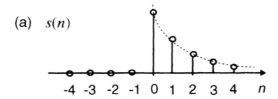

(b) $s_e(n)$

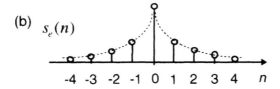

(c) $s_o(n)$

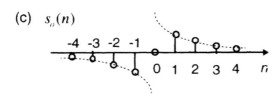

Figure 4.4.1 Decomposition of a real causal sequence into $s_e(n)$ and $s_o(n)$, respectively.

and

$$s_o(n) = \begin{cases} s_e(n) & \text{for } n > 0 \\ -s_e(n) & \text{for } n < 0 \\ 0 & \text{for } n = 0. \end{cases}$$

Consequently, for given $s_o(n)$ and $s(0)$, $s_e(n)$ can be obtained and conversely $s_o(n)$ can be found from $s_e(n)$. Therefore the transform $S(e^{j\omega})$ can be obtained from $S_R(e^{j\omega})$ or $S_I(e^{j\omega})$.

Given $S_R(e^{j\omega})$, try to find $S_I(e^{j\omega})$. Taking the inverse Fourier transform of $S_R(e^{j\omega})$, we get the even component $s_e(n)$ as follows:

$$s_e(n) = \frac{1}{2\pi} \int_{-\pi}^{\pi} S_R(e^{j\omega}) \, d\omega.$$

The odd sequence $s_o(n)$ can be expressed in terms of $s_e(n)$ as

$$s_o(n) = \begin{cases} s_e(n) & \text{for } n > 0 \\ 0 & \text{for } n = 0 \\ -s_e(n) & \text{for } n < 0. \end{cases}$$

Finally, by taking the Fourier transform of $s_o(n)$, $S_I(e^{j\omega})$ can be obtained as

$$-jS_I(e^{j\omega}) = \sum_{n=-\infty}^{\infty} s_o(n) e^{-j\omega n}.$$

In the remaining part of this section we show that we can derive $S_I(e^{j\omega})$ from $S_R(e^{j\omega})$ using the example of a sequence. Consider a real and causal sequence $s(n)$ with a real constant a ($|a| < 1$):

$$s(n) = \begin{cases} a^n & n \geq 0 \\ 0 & n < 0. \end{cases}$$

The Fourier transform of $s(n)$ is

$$S(e^{j\omega}) = \sum_{n=0}^{\infty} a^n e^{-j\omega n} = \frac{1}{1 - ae^{-j\omega}}$$

$$= \frac{1 - a\cos\omega}{1 - 2a\cos\omega + a^2} - j\frac{a\sin\omega}{1 - 2a\cos\omega + a^2}.$$

Therefore, if

$$S(e^{j\omega}) = S_R(e^{j\omega}) - jS_I(e^{j\omega}),$$

then

$$S_R(e^{j\omega}) = \frac{1 - a\cos\omega}{1 - 2a\cos\omega + a^2}$$

and

$$S_I(e^{j\omega}) = \frac{a\sin\omega}{1 - 2a\cos\omega + a^2}.$$

Now let us derive $S_I(e^{j\omega})$ from $S_R(e^{j\omega})$. $S_R(e^{j\omega})$ can be divided into the following partial fraction expansion:

$$S_R(e^{j\omega}) = \frac{1 - a\cos\omega}{1 - 2a\cos\omega + a^2} = \frac{2 - a(e^{j\omega} + e^{-j\omega})}{1 - a(e^{j\omega} + e^{-j\omega}) + a^2}\frac{1}{2}$$

$$= \frac{2 - ae^{j\omega} - ae^{-j\omega}}{(1 - ae^{-j\omega})(1 - ae^{j\omega})}\frac{1}{2} = \frac{\frac{1}{2}}{1 - ae^{-j\omega}} + \frac{\frac{1}{2}}{1 - ae^{j\omega}}.$$

Since $|a| < 1$,

$$S_R(e^{j\omega}) = \tfrac{1}{2}(1 + ae^{-j\omega} + a^2e^{-j2\omega} + \cdots)$$
$$+ \tfrac{1}{2}(1 + ae^{j\omega} + a^2e^{j2\omega} + \cdots)$$

$$= \tfrac{1}{2}\sum_{n=0}^{\infty} a^n e^{-j\omega n} + \tfrac{1}{2}\sum_{n=0}^{\infty} a^n e^{j\omega n}$$

$$= \tfrac{1}{2}\sum_{n=0}^{\infty} a^n e^{-j\omega n} + \tfrac{1}{2}\sum_{n=-\infty}^{-\infty} a^{-n} e^{-j\omega n}.$$

From the definition of the Fourier transform of sequences given in Chapter 2, the first and second terms in the final expression of the equation above are the Fourier representations for sequences. Thus we have an even sequence, i.e.,

$$s_e(n) = \begin{cases} \tfrac{1}{2}a^n & \text{for } n > 0 \\ 1 & \text{for } n = 0 \\ \tfrac{1}{2}a^{-n} & \text{for } n < 0. \end{cases}$$

The odd sequence component is obtained as follows:

$$s_o(n) = \begin{cases} s_e(n) = \tfrac{1}{2}a^n & \text{for } n > 0 \\ 0 & \text{for } n = 0 \\ -s_e(n) = -\tfrac{1}{2}a^{-n} & \text{for } n < 0. \end{cases}$$

Therefore, through Fourier transform of $s_o(n)$, we can get $S_I(e^{j\omega})$ as follows:

$$-jS_I(e^{j\omega}) = \sum_{n=-\infty}^{\infty} s_o(n)e^{-j\omega n} = \frac{1}{2}\sum_{n=0}^{\infty} a^n e^{-j\omega n} - \frac{1}{2}\sum_{n=-\infty}^{0} a^{-n} e^{-j\omega n}$$

$$= \frac{1}{2}\left\{ \sum_{n=0}^{\infty} a^n e^{-j\omega n} - \sum_{n=0}^{\infty} a^n e^{j\omega n} \right\}$$

$$= \frac{1}{2}\left\{ \frac{1}{1 - ae^{-j\omega}} - \frac{1}{1 - ae^{j\omega}} \right\}$$

$$= -j\frac{a \sin \omega}{1 - 2a \cos \omega + a^2},$$

or equivalently

$$S_I(e^{j\omega}) = \frac{a \sin \omega}{1 - 2a \cos \omega + a^2}.$$

The result is what we expected.

4.4.3   Minimum-phase Sequences

Consider minimum-phase sequences or systems. The frequency response function of a system is represented in the form

$$H(e^{j\omega}) = |H(e^{j\omega})|e^{j\theta(e^{j\omega})}$$

as stated in Section 4.1.2. Taking the natural complex logarithm of both sides of the equation, we get

$$\ln H(e^{j\omega}) = \ln |H(e^{j\omega})| + j\theta(e^{j\omega}).$$

The right-hand side of this equation consists of real and imaginary parts. If the frequency response function $\ln H(e^{j\omega})$ is the Fourier transform of a real and causal sequence $c(n)$, then we could derive the real part (log magnitude in nepers) from the imaginary part (phase function in radians), as already described in this section. This means that the phase function can be obtained from the log magnitude in nepers and vice versa.

Not all systems or sequences have this property. It is only the case where the sequence $c(n)$ which is the inverse Fourier transform of $\ln H(e^{j\omega})$,

$$c(n) = \frac{1}{2\pi} \int_{-\pi}^{\pi} \ln H(e^{j\omega})e^{j\omega}\,d\omega,$$

is real and a causal sequence. In what conditions will $c(n)$ be real and causal? First, confirm that the complex logarithm $\ln H(e^{j\omega})$ is always real. From the relationship discussed in Section 4.1.2, the magnitude can be written as

$$|H(e^{j\omega})| = \sqrt{H_R^2(e^{j\omega}) + H_I^2(e^{j\omega})},$$

where $|H(e^{j\omega})|$ is seen to be an even function of ω, because $H_R(e^{j\omega})$ is a real and even function and, although $H_I(e^{j\omega})$ is an real and odd function, the square of $H_I(e^{j\omega})$ is an even function. Therefore, the inverse Fourier transform of $|H(e^{j\omega})|$ is real.

From the discussion in Section 4.1.2, we have

$$\theta(e^{j\omega}) = \tan^{-1}\frac{H_I(e^{j\omega})}{H_R(e^{j\omega})},$$

where $\theta(e^{j\omega})$ is an odd function in ω, because $\tan^{-1}$ and $H_I(e^{j\omega})/H_R(e^{j\omega})$ are both odd functions. The inverse Fourier transform of the real and odd function $\theta(e^{j\omega})$ is always pure imaginary because

$$\frac{1}{2\pi}\int_{-\pi}^{\pi}\theta(e^{j\omega})e^{j\omega n}\,d\omega = \frac{j}{2\pi}\int_{-\pi}^{\pi}\theta(e^{j\omega})\sin\omega n\,d\omega.$$

Therefore, the inverse Fourier transform results in a real number:

$$c(n) = \frac{1}{2\pi} \int_{-\pi}^{\pi} \ln H(e^{j\omega}) e^{j\omega n} \, d\omega$$

$$= \frac{1}{2\pi} \int_{-\pi}^{\pi} \left\{ \ln |H(e^{j\omega})| + j\theta(e^{j\omega}) \right\} e^{j\omega n} \, d\omega.$$

Next consider the causality of $c(n)$. Assume that $H(e^{j\omega})$ is the frequency response of a system whose transfer function $H(z)$ is described by a rational function of z^{-1} as shown in Section 4.2.1,

$$\ln H(z) = \ln \frac{b_0 \prod_{l=1}^{L-1} (1 - d_l z^{-1})}{\prod_{k=1}^{K-1} (1 - c_k z^{-1})}$$

$$= \ln b_0 + \sum_{l=1}^{L-1} \ln(1 - d_l z^{-1}) - \sum_{k=1}^{K-1} \ln(1 - c_k z^{-1}),$$

where the d_l are the zeros, and the c_k are the poles of $H(z)$, and $H(z)$ consists of a sum of terms of identical form except for $\ln b_0$. To examine the causality of the system it is sufficient to consider the causality of the sequence corresponding to the term $\ln(1 - az^{-1})$, where a may be any complex number.

Now consider the inverse z-transform of

$$\ln(1 - az^{-1}).$$

We have a power series expansion formula as

$$\ln(1 - x) = -\sum_{n=1}^{\infty} \frac{x^n}{n} \qquad |x| < 1.$$

Applying this expansion formula to $\ln(1 - az^{-1})$, we get a power series in z^{-1} when $|az^{-1}| < 1$, or equivalently $|z| > |a|$:

$$\ln(1 - az^{-1}) = -\sum_{n=1}^{\infty} \frac{a^n}{n} z^{-n}.$$

When $|az^{-1}| > 1$, or equivalently $|z| < |a|$, similarly we get

$$\ln(1 - az^{-1}) = \ln\left((az^{-1})\frac{1 - az^{-1}}{az^{-1}}\right) = \ln\left((-az^{-1})\frac{z - a}{-a}\right)$$

$$= \ln\left((-az^{-1})\frac{a - z}{a}\right) = \ln[(-az^{-1})(1 - za^{-1})]$$

$$= \ln(-az^{-1}) + \ln(1 - a^{-1}z) = \ln(-az^{-1}) - \sum_{n=-1}^{-\infty} \frac{a^n}{n}z^{-n}$$

$$= \ln(-a) + \ln z^{-1} - \sum_{n=-1}^{-\infty} \frac{a^n}{n}z^{-n}.$$

The power series expansion of $\ln(1 - az^{-1})$ above gives us the inverse Fourier transform $c(n)$ of $\ln(1 - ae^{-j\omega})$ as follows. If $|a| < |z| = 1$, then

$$c(n) = \begin{cases} -\dfrac{a^n}{n} & n > 0 \\ 0 & n \leq 0, \end{cases}$$

and if $|a| > |z| = 1$, then

$$c(n) = \begin{cases} -\dfrac{\cos(n\pi)}{n} & n > 0 \\ \dfrac{a^n}{n} - \dfrac{\cos(n\pi)}{n} & n < 0 \\ \ln(-a) & n = 0. \end{cases}$$

The terms appearing in the equation immediately above,

$$-\frac{\cos(n\pi)}{n} \qquad \text{and} \qquad \ln(-a),$$

are the inverse Fourier transform of

$$\ln z^{-1}\big|_{z=e^{j\omega}} = -j\omega \qquad -\pi \leq \omega < \pi$$

and the constant $\ln(-a)$, respectively, where the inverse Fourier transform can be written as

$$\frac{1}{2\pi}\int_{-\pi}^{+\pi} -j\omega e^{j\omega n}\, d\omega = \frac{-j}{2\pi}\left[\omega \frac{e^{j\omega n}}{jn}\right]_{-\pi}^{+\pi}$$

$$= \frac{-1}{2\pi n}\{\pi e^{jn\pi} + \pi e^{-jn\pi}\} = \frac{-1}{n}\cos(n\pi).$$

We see that the sequence $c(n)$ is causal only for $|a| < 1$, i.e., a is inside the unit circle.

This discussion is valid for the zeros and poles of all other terms in $\ln H(z)$. Therefore, if the poles and zeros constructing $H(z)$ are all inside the unit circle, the sequence $c(n)$ derived by the inverse Fourier transform of $\ln H(e^{j\omega})$ is causal. Systems satisfying this condition are called 'minimum-phase' systems. The log magnitude response and the phase response of a minimum-phase system are derived from each other. The relation is referred to as Hilbert transform. The sequence $c(n)$ obtained above is called the 'complex cepstrum' which will be discussed again in Chapter 8 using continuous function forms.

4.5   Minimum-phase and All-pass Systems

4.5.1   Minimum-phase Systems

In this section important and interesting features of transfer functions are presented. Consider two single-zero systems of which the transfer functions are

$$H_1(z) = 1 - az^{-1}$$

and

$$H_2(z) = z^{-1} - a^*,$$

where a is a complex number and $*$ indicates taking the complex conjugate. The pole–zero plots of the two systems are depicted in Fig. 4.5.1, where $|a| < 1$ is assumed and $H_1(z)$ has one pole and one zero inside the unit circle and $H_2(z)$ has one pole inside and one zero outside the unit circle. The zero of $H_2(z)$ exists at the reciprocal conjugate location of the zero of $H_1(z)$.

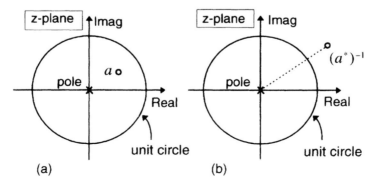

Figure 4.5.1   Pole–zero plots showing (a) location of zero a and (b) the reciprocal conjugate position of a.

Let us compare the magnitudes of $H_1(z)$ and $H_2(z)$ on the unit circle. We can see

$$|H_1(z)| = |H_2(z)| \qquad \text{for } z = e^{j\omega},$$

or equivalently,

$$|1 - ae^{-j\omega}| = |e^{-j\omega}(e^{j\omega} - a)| = |e^{j\omega} - a| = |e^{-j\omega} - a^*|.$$

But the phase responses which are obtained by evaluating $H_1(z)$ and $H_2(z)$ on the unit circle are different. Indeed,

$$H_1(z)\big|_{z=e^{j\omega}} = H_1(e^{j\omega}) = 1 - ae^{-j\omega}$$

and

$$H_2(z)\big|_{z=e^{j\omega}} = e^{-j\omega} - a^* = e^{-j\omega}(1 - a^* e^{j\omega})$$

$$= e^{-j\omega}(1 - ae^{-j\omega})^* = e^{-j\omega}H_1^*(e^{j\omega}).$$

The phase responses are shown in Fig. 4.5.2. The two phase functions are actually different.

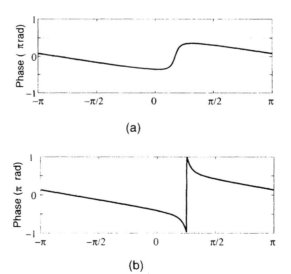

(a)

(b)

Figure 4.5.2   The phase responses of two systems: (a) $1 - az^{-1}$; (b) $1 - a^{-1*}z^{-1}$.

The above discussion suggests that there could be transfer functions of which the magnitude responses are identical even though their phase responses are different. In other words, the zeros of a transfer function could be moved to the conjugate reciprocal positions on the z-plane, keeping the magnitude response unchanged. Also, when the magnitude response is given, we cannot take arbitrary phase responses. If the transfer function has N zeros on the entire z-plane, we can place any number of zeros ($\leq N$) inside the unit circle and the remaining zeros outside the unit circle, so that the number of possible phase responses will be 2^N cases for a fixed magnitude response.

The condition that all the zeros and the poles are inside the unit circle (all the poles must be inside the unit circle for any stable system) is referred to as minimum-phase (condition); and the condition that all the zeros are outside is called maximum-phase (condition). The terminology of 'minimum-phase' comes from the fact that the phase response in the minimum-phase condition varies in the smallest range as ω changes from $-\pi$ to π among other zero locations keeping the magnitude unchanged. The maximum-phase response varies in the largest range among the 2^N phases.

Example 4.5.1

Examine the phase response of a system with transfer function composed of two zeros,

$$H(z) = (1 - az^{-1})(1 - bz^{-1}),$$

where a and b are zeros. Depending on the positions of a and b, the system shows four different phase responses. This is shown in Fig. 4.5.3 together with the magnitude response, where the phase values are expressed by the principal values. The phase curve in part (a) of Fig. 4.5.3 which corresponds to the minimum-phase condition varies in the smallest region compared with the remaining three phase responses. If we unwrap the phase curves (b) through (d), the phase changes are 4π, 2π and 2π, respectively, while the minimum-phase curve (a) changes about π.

From a time sequence point of view, different phase responses produce different sequences. The minimum-phase sequence is likely to be concentrated toward the time origin. For example, if $s_{\min}(n)$, is a minimum-phase sequence with the transfer function $S_{\min}(z)$, then the inequality

$$|s_{\min}(0)| > |s(0)|$$

holds where $s(0)$ is the initial value of a non-minimum-phase sequence of which the transfer function equals $S(z)$. In the later part of this section we will discuss an all-pass system. The inequality shown above will be clarified.

In Fig. 4.5.4 are illustrated the pole–zero plots and the impulse response sequences corresponding to each pole–zero plot for a two-zero transfer

104   *Fundamentals of Acoustic Signal Processing*

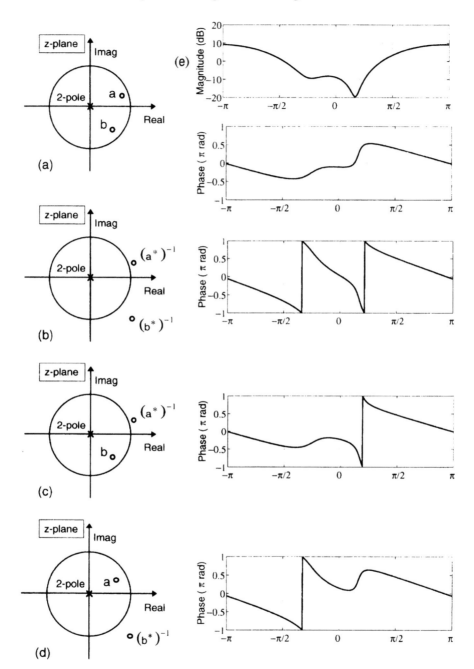

Figure 4.5.3   The phase and magnitude response of the system $H(z) = (1 - az^{-1})$ $(1 - bz^{-1})$. In (a)–(d) four different phase responses corresponding to the pole–zero plots are shown; (e) the magnitude response of the system.

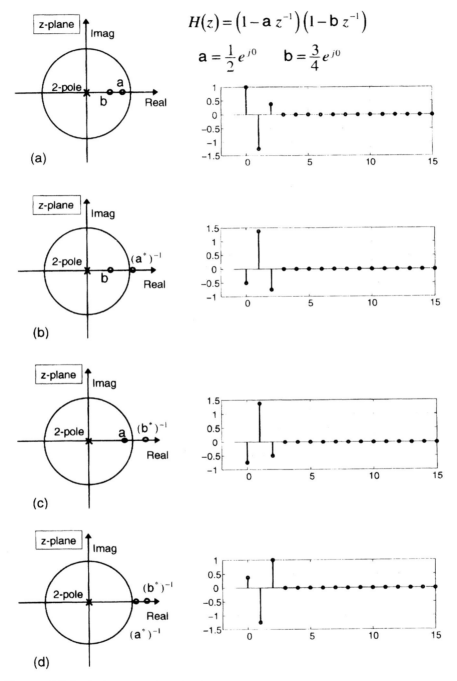

Figure 4.5.4   Pole–zero plots for four pole–zero allocations and their impulse responses.

function

$$H(z) = (1 - az^{-1})(1 - bz^{-1})$$

for $a = \frac{1}{2}$ and $b = \frac{3}{4}$. Figure 4.5.4(a) shows the minimum-phase condition where the initial value $h(0)$ can be seen to be the largest.

4.5.2   All-pass Systems

We proceed to discussion of all-pass systems or sequences. Consider a transfer function $H_{ap}(z)$ defined by $H_2(z)$ divided by $H_1(z)$:

$$H_{ap}(z) = \frac{H_2(z)}{H_1(z)} = \frac{z^{-1} - a^*}{1 - az^{-1}},$$

where $H_2(z)$ and $H_1(z)$ are as introduced in Section 4.5.1. If the magnitude of $H_{ap}(z)$ is evaluated on the unit circle, the equation

$$|H_{ap}(e^{j\omega})| = \left| \frac{e^{-j\omega} - a^*}{1 - ae^{-j\omega}} \right| = \left| \frac{e^{-j\omega} - a^*}{e^{-j\omega}(e^{j\omega} - a)} \right| = \frac{|e^{-j\omega} - a^*|}{|e^{-j\omega}(e^{j\omega} - a)|}$$

$$= \frac{|e^{-j\omega} - a^*|}{|e^{j\omega} - a|} = \frac{|e^{-j\omega} - a^*|}{|e^{-j\omega} - a^*|} = 1$$

can be derived. $H_{ap}(z)$ is referred to as an *all-pass system* or *filter*, because the magnitude of $H_{ap}(e^{j\omega})$ is unity for all frequencies ω. But $H_{ap}(z)$ has its own phase response.

Using the property of all-pass systems, any transfer function can be expressed by the equation

$$H(z) = H_{min}(z) \cdot H_{ap}(z).$$

A cascade connection of any number of all-pass systems of the form indicated again makes an all-pass system. Consider a transfer function as

$$H(z) = \frac{b_0 \displaystyle\prod_{l=0}^{L}(1 - d_l z^{-1})}{\displaystyle\prod_{k=0}^{K-1}(1 - c_k z^{-1})}.$$

If it has several zeros outside the unit circle, then we can reflect the zeros inside the unit circle as follows. Let one outside zero of $H(z)$ be a_0^{-1} (for $|a_0| < 1$). $H(z)$ can then be written as

$$H(z) = K(z)(z^{-1} - a_0)\frac{1 - a_0^* z^{-1}}{1 - a_0^* z^{-1}} = K(z)(1 - a_0^* z^{-1})\frac{z^{-1} - a_0}{1 - a_0^* z^{-1}},$$

where $K(z)$ indicates the remainder part of $H(z)$ after deleting the zero term $(z^{-1} - a_0)$. The third term of the right-hand side,

$$\frac{z^{-1} - a_0}{1 - a_0^* z^{-1}},$$

is an all-pass transfer function as already shown.

By this procedure we can reflect a zero (a_0^{-1}) located outside the unit circle inside the unit circle. Continuing the zero reflection procedure for all the zeros outside the unit circle, finally $H(z)$ can be changed to the form

$$H(z) = H_{min}(z) \cdot H_{ap}(z).$$

Since the zero reflection procedure does not cause any change in the magnitudes $|H(e^{j\omega})|$ and $|H_{ap}(e^{j\omega})| = 1$, we have

$$|H(e^{j\omega})| = |H_{min}(e^{j\omega})H_{ap}(e^{j\omega})| = |H_{min}(e^{j\omega})|.$$

Making use of

$$S(z) = S_{min}(z) \cdot S_{ap}(z),$$

we can show the inequality previously mentioned,

$$|s_{min}(0)| > |s(0)|.$$

Using the initial value theorem for causal sequences presented in Section 3.4.5, we get

$$\lim_{z \to \infty} S(z) = s(0),$$

$$\lim_{z \to \infty} S_{min}(z) = s_{min}(0),$$

$$\lim_{z \to \infty} S_{ap}(z) = s_{ap}(0).$$

Furthermore, since an all-pass transfer function of the form

$$S_{ap}(z) = \frac{z^{-1} - a}{1 - a^* z^{-1}}$$

converges to a finite value $-a$ with $z \to \infty$ and a is less than unity,

$$\left| \lim_{z \to \infty} S_{ap}(z) \right| = |s_{ap}(0)| = |a| < 1$$

holds well. The inequality is correct for an all-pass system obtained by the cascade connection of any number of all-pass systems of the form

$$S_{ap}(z) = \frac{z^{-1} - a}{1 - a^* z^{-1}},$$

because $|a| < 1$. This leads to the inequality as

$$|s(0)| = |\lim_{z \to \infty} S(z)| = |\lim_{z \to \infty} S_{min}(z) \cdot S_{ap}(z)|$$

$$= |s_{min}(0)| \cdot |s_{ap}(0)| < |s_{min}(0)|.$$

This shows what we want to derive.

4.6   Inverse Filter

The inverse filter of a stable and causal sequence $h(n)$ is defined by the sequence $g(n)$ satisfying the equation

$$h(n) * g(n) = \delta(n),$$

where $*$ indicates taking a convolution and $\delta(n)$ is a unit impulse at $n = 0$. Figure 4.6.1 illustrates the schematic diagram explaining the notion of inverse filtering. It shows that the convolution sum of $h(n)$ and $g(n)$ becomes a unit impulse. The inverse filter problem is the procedure for finding a stable and causal sequence $g(n)$ when $h(n)$ is given.

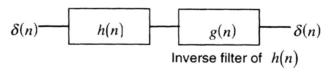

Inverse filter of $h(n)$

Figure 4.6.1   Schematic diagram explaining an inverse filter.

Let $H(z)$ and $G(z)$ be z-transforms of $h(n)$ and $g(n)$, respectively. We get

$$H(z) \cdot G(z) = 1.$$

Looking at this relation, $G(z)$ seems to be obtained by taking the reciprocal of $H(z)$ as

$$G(z) = \frac{1}{H(z)}.$$

However, this method cannot always resolve inverse problems, since the zeros of $H(z)$ are not guaranteed to be inside the unit circle on the z-plane.

The zeros of $H(z)$ will in turn become poles of $G(z)$. A pole outside the unit circle leads to an unstable sequence.

The inverse filter of $H(z)$ can be obtained if and only if the zeros and poles are all inside the unit circle. In other words, the inverse filter exists only when $H(z)$ is of minimum phase. If we can obtain the inverse filter, the inverse filter is always of minimum phase.

Example 4.6.1

Consider $H(z)$ given below and find the inverse filter of $H(z)$:

$$H(z) = 1 - az^{-1},$$

where a is real and less than 1. $H(z)$ is of minimum-phase so that we can find the inverse $G(z)$ by taking the reciprocal:

$$G(z) = \frac{1}{H(z)} = \frac{1}{1 - az^{-1}}.$$

Since $|a| < 1$, $G(z)$ is stable. Figure 4.6.2 shows the impulse response sequences of $H(z)$ and $G(z)$ for $a = 0.8$. If $|a| > 1$, then $G(z)$ does not exist, because the zero of $H(z)$ is outside the unit circle.

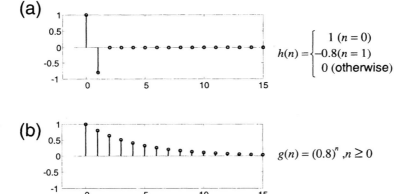

(a)

$$h(n) = \begin{cases} 1 \ (n = 0) \\ -0.8(n = 1) \\ 0 \ (\text{otherwise}) \end{cases}$$

(b)

$$g(n) = (0.8)^n, n \geq 0$$

Figure 4.6.2   An example of the inverse filter impulse response. (a) The impulse response $h(n)$ and (b) the inverse filter impulse response $g(n)$.

In many cases, impulse response sequences measured in acoustic fields are not of minimum-phase, so that exact inverse filters cannot be obtained. In practical applications, an approximation method for finding the required inverse systems has been developed. The details of these will be presented in Chapter 9.

4.7   Summary

In this chapter we described transfer functions and frequency response functions of linear time-invariant systems. First we showed the frequency response of a system obtained by the input of $e^{j\omega_0 n}$ and the linear combination of $e^{j\omega_0 n}$. Next we discussed the transfer function of a system represented by a linear constant coefficient difference equation and we presented a pole–zero plot of a transfer function on the z-plane and the geometrical interpretation of the magnitude and phase response. Properties of linear system frequency responses are presented particularly in terms of the stability and causality. The relationships between the real and imaginary parts of frequency response functions of real sequences are also described. Finally, we presented the decomposition of a transfer function into an all-pass and a minimum-phase system and described the basic idea of inverse filters.

5 Discrete Fourier Transform

5.0 Introduction

In the preceeding chapters we discussed discrete-time sequences and their Fourier transform representations expressed by continuous normalized frequency. This is called *discrete-time Fourier transformation* in order to distinguish it from the discrete Fourier transform to be presented later in this chapter. That is, while sequences are of discrete-time expression, frequency representations of sequences and frequency responses of systems are expressed by functions of a continuous frequency variable. Since the continuous-frequency representations are inconvenient for implementation by digital computer software or hardware, the frequency variable is also desired to be discrete. As discussed in Chapter 2 a continuous-time signal to be sampled is frequency band-limited and its Fourier transform is periodic in ω with period 2π, and therefore a frequency representation of a sequence may be composed of a finite number of frequency components.

In this chapter we discuss the Fourier representations for finite-length time sequences, referred to as the discrete Fourier transform (DFT), which is not a continuous-frequency function but a discrete-frequency sequence obtained by sampling the discrete-time Fourier transform of the sequence with an equally spaced frequency interval. We show some relationships between DFT and the Fourier series representation of periodic sequences, some properties of DFT, efficient algorithms for computing DFT, and applications of DFT to the convolution computations.

5.1 Discrete Fourier Representation of Finite-length Sequences

Consider a finite-length time sequence $s(n)$ $(n = 0, 1, \ldots, N-1)$. As stated in Section 2.3.2, $s(n)$ can be represented by the discrete-time Fourier transform representation

$$s(n) = \frac{1}{2\pi} \int_{-\pi}^{\pi} S_d(e^{j\omega}) e^{j\omega n} \, d\omega,$$

where

$$S_d(e^{j\omega}) = \sum_{n=-\infty}^{\infty} s(n)e^{-j\omega n}.$$

In this section we discuss the discrete Fourier transform (DFT) representation of finite-length sequences in which a discrete frequency variable is used instead of continuous frequency. Although the sequence $s(n)$ is of finite-length N, it should be noted that both the discrete-time Fourier transform and the DFT are periodic with period 2π and N, respectively.

Since $S_d(e^{j\omega})$ is a periodic function of ω with period 2π, we try to sample $S_d(e^{j\omega})$ at M equally spaced frequency points over the interval $-\pi \leq \omega < \pi$, i.e., $\omega_m = (2\pi/M)m$ with m as an integer from 0 to $M-1$. We can then get a sequence $S(m)$ in the frequency domain from the discrete-time Fourier transform $S_d(e^{j\omega})$ for a finite-length sequence $s(n)$ as

$$S(m) = S_d(e^{j(2\pi/M)m}) = \sum_{n=0}^{N-1} s(n)e^{-j(2\pi/M)mn} \qquad 0 \leq m \leq M-1.$$

If we can determine how large an M is necessary to reconstruct the finite-length sequence $s(n)$ exactly, the continuous-frequency function $S_d(e^{j\omega})$ can be represented by an M-sample sequence.

Since $S(m)$ represents a coefficient of the frequency component with frequency $(2\pi/M)m$, the sequence to be reconstructed, say $\hat{s}(n)$, can be represented by

$$\hat{s}(n) = \sum_{m=0}^{M-1} S(m)e^{j(2\pi/M)mn} \qquad 0 \leq n \leq N-1,$$

where the period of $\hat{s}(n)$ is M, if we extend n outside the interval above. Substitution of

$$S(m) = \sum_{k=0}^{N-1} s(k)e^{-j(2\pi/M)mk}$$

into this equation leads to

$$\hat{s}(n) = \sum_{m=0}^{M-1} \left\{ \sum_{k=0}^{N-1} s(k)e^{-j(2\pi/M)mk} \right\} e^{j(2\pi/M)mn}.$$

Interchanging the two summations, we obtain

$$\hat{s}(n) = \sum_{k=0}^{N-1} s(k) \sum_{m=0}^{M-1} e^{j(2\pi/M)(n-k)m} = M \sum_{k=0}^{N-1} s(k)\delta(n - k \pm Ml),$$

where l is a positive integer and we use the relation as

$$\sum_{m=0}^{M-1} e^{j(2\pi/M)(n-k)m} = \frac{1 - \{e^{j(2\pi/M)(n-k)}\}^M}{1 - e^{j(2\pi/M)(n-k)}} = 0 \qquad \text{for } k \neq n \pm Ml.$$

If $M < N$, then

$$\hat{s}(n) = \hat{s}(n \pm Ml) = M[s(n) + s(n \pm M) + \cdots + s(n \pm Ml)],$$

where $0 \leq n \pm Ml \leq N - 1$, and l is a positive integer. This implies that $\hat{s}(n)$ cannot be determined for $M \leq n \leq N - 1$.

Let us consider an example where $M = 5$ and $N = 8$. The recovered sequence $\hat{s}(n)$ becomes

$$\hat{s}(0) = 5[s(0) + s(0 + 5)] = 5[s(0) + s(5)] \neq 5s(0),$$
$$\hat{s}(1) = 5[s(1) + s(1 + 5)] = 5[s(1) + s(6)] \neq 5s(1),$$
$$\hat{s}(2) = 5[s(2) + s(2 + 5)] = 5[s(2) + s(7)] \neq 5s(2),$$
$$\hat{s}(3) = 5[s(3)],$$
$$\hat{s}(4) = 5[s(4)],$$
$$\hat{s}(5) = 5[s(5) + s(5 - 5)] = 5[s(5) + s(0)] = \hat{s}(0) \neq 5s(5),$$
$$\hat{s}(6) = 5[s(6) + s(6 - 5)] = 5[s(6) + s(1)] = \hat{s}(1) \neq 5s(6),$$
$$\hat{s}(7) = 5[s(7) + s(7 - 5)] = 5[s(7) + s(2)] = \hat{s}(2) \neq 5s(7).$$

We can see that the reconstructed sequence has period $M = 5$ and the original sequence is not exactly reconstructed.

If $M \geq N$, we get in the same way to the case $M < N$:

$$\hat{s}(n) = \hat{s}(n + Ml) = \begin{cases} Ns(n) & 0 \leq n \leq N - 1 \\ 0 & N \leq n \leq M - 1, \end{cases}$$

where i is a positive integer. The original sequence can be reconstructed exactly by dividing by N in the interval $0 \leq n \leq N - 1$.

To summarize, the original sequence $s(n)$ can be exactly recovered by the sum of $M(\geq N)$ frequency components, while $s(n)$ cannot be recovered by the sum of M less than N frequency components. Thus the discrete

Fourier transform (DFT) of the sequence $s(n)$ of length N is defined as

$$S(m) = \sum_{n=0}^{N-1} s(n)e^{-j(2\pi/N)mn} \qquad 0 \le m \le N-1$$

and

$$s(n) = \frac{1}{N} \sum_{m=0}^{N-1} S(m)e^{j(2\pi/N)mn} \qquad 0 \le n \le N-1.$$

This shows that an N-length time sequence and an N-length frequency sequence are interexchangeable. The two sequences are called a *DFT pair*.

In computations of DFT, it is often convenient to use length L rather than length N of the given sequence, where $L > N$, in order to exploit more efficient computing algorithms called fast Fourier transform (FFT) algorithms, to be discussed in Section 5.4. In this case the DFT could be carried out for the L-length sequence $s_L(n)$ produced by attaching an $(L - N)$-length zero sequence (the procedure is called 'zero padding') at the end of $s(n)$ such as

$$s_L(n) = \begin{cases} s(n) & 0 \le n \le N-1 \\ 0 & N \le n \le L-1. \end{cases}$$

The DFT of the augmented sequence $s_L(n)$ is

$$S_L(m) = \sum_{n=0}^{L-1} s_L(n)e^{-j(2\pi/L)nm} \qquad 0 \le n \le L-1.$$

By this transform we can get an L-point DFT sequence; and by inverse DFT the sequence $s_L(n)$ can be recovered exactly.

Example 5.1.1

Consider the DFT representation of the time sequence of length 8:

$$s(n) = a^n \qquad 0 \le n \le 7.$$

The DFT of $s(n)$ is

$$S(m) = \sum_{n=0}^{7} a^n e^{-j(2\pi/8)nm} = \sum_{n=0}^{7} (ae^{-j(\pi/4)m})^n$$

$$= \frac{1 - a^8 e^{-j2\pi m}}{1 - ae^{-j(\pi/4)m}} = \frac{1 - a^8}{1 - ae^{-j(\pi/4)m}},$$

where $0 \leq m \leq 7$. The reconstructed time sequence $\hat{s}(n)$ is

$$\hat{s}(n) = \sum_{m=0}^{7} S(m)e^{j(2\pi/8)mn} = \sum_{m=0}^{7} \frac{1-a^8}{1-ae^{-j(\pi/4)m}} \cdot e^{j(2\pi/8)mn}$$

$$= \sum_{m=0}^{7} \left\{ \sum_{k=0}^{7} a^k e^{-j(2\pi/8)mk} \right\} e^{j(2\pi/8)mn}$$

$$= \sum_{k=0}^{7} a^k \sum_{m=0}^{7} e^{j(2\pi/8)(n-k)m} = 8 \sum_{k=0}^{7} a^k \delta(n-k) = 8a^n = 8s(n),$$

where $0 \leq n \leq 7$. The sequences $s(n)$ and $S(m)$ are illustrated in Fig. 5.1.1 for $a = 0.9$. It is clear that both the discrete-time Fourier transform $S_d(e^{j\omega})$ and the DFT of the sequence $s(n)$ are periodic and that the DFT sequence consists of the sample values obtained by sampling the

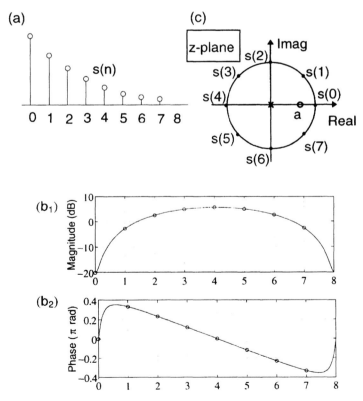

Figure 5.1.1 (a) Exponential sequence of length 8. (b) Magnitude and phase. (c) The sampling points on the unit circle.

discrete-time Fourier transform function of the continuous variable ω at equally spaced discrete frequency points $(2\pi/N)k$ (in this example, $N = 8$ and $k = 0, 1, \ldots, N-1$).

Example 5.1.2

In the previous example, consider the case where the number of frequency sampling points $M(= 10)$ is larger than the time sequence length $N(= 8)$. Then the DFT is

$$S(m) = \sum_{n=0}^{N-1} s(n)e^{-j(2\pi/M)nm} = \sum_{n=0}^{7} a^n e^{-j(2\pi/10)nm}$$

$$= \sum_{n=0}^{7} (ae^{-j(\pi/5)m})^n = \frac{1 - (ae^{-j(\pi/5)m})^8}{1 - ae^{-j(\pi/5)m}},$$

where $0 \le m \le M-1$. The reconstructed time sequence $\hat{s}(n)$ is obtained as

$$\hat{s}(n) = \sum_{m=0}^{9} S(m)e^{j(2\pi/10)mn} = \sum_{m=0}^{9} \frac{1 - (ae^{-j(2\pi/10)m})^8}{1 - ae^{-j(2\pi/10)m}} e^{j(2\pi/10)mn}$$

$$= \sum_{m=0}^{9} \left\{ \sum_{k=0}^{7} a^k e^{-j(2\pi/10)mk} \right\} e^{j(2\pi/10)mn}$$

$$= \sum_{k=0}^{7} a^k \sum_{m=0}^{9} e^{j(2\pi/10)(n-k)m} = 10 \sum_{k=0}^{7} a^k \delta(n-k)$$

$$= \begin{cases} 10s(n) & 0 \le n \le 7 \\ 0 & 8 \le n \le 9. \end{cases}$$

The sequences $\hat{s}(n)$ and $S(m)$ are illustrated in Fig. 5.1.2. The difference between $\hat{s}(n)$ and $s(n)$ is a newly generated length-2 zero sequence appearing at the end of $\hat{s}(n)$. So the total length of the reconstructed sequence $\hat{s}(n)$ will be 10 and $\hat{s}(n)$ has a period 10 as depicted in Fig. 5.1.2.

5.2   DFT and Discrete Fourier Series (DFS)

In this section the relationship between DFT and discrete Fourier series (DFS) representations of sequences is described. Consider a periodic sequence $\tilde{s}(n)$ (henceforth, we use a tilde ($\sim$) on the top of $s(n)$ to indicate

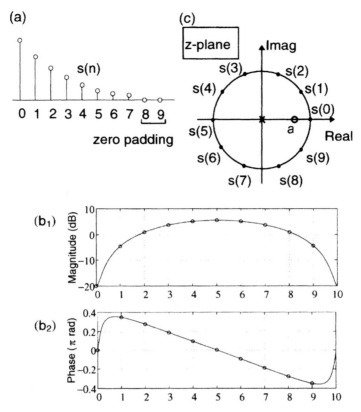

Figure 5.1.2 (a) Exponential sequence of length 8 with zero padding of length 2. (b) Magnitude and phase. (c) The sampling points on the unit circle.

a periodic sequence) with period N, i.e.,

$$\tilde{s}(n) = \tilde{s}(n + lN) \qquad l \text{ integer.}$$

The periodic sequence $\tilde{s}(n)$ does not have a Fourier transform because $\tilde{s}(n)$ is not absolutely summable, although we formally introduced a Fourier representation of a periodic sequence $x(n)$ using an infinite series as

$$x(n) = \sum_{m=-\infty}^{+\infty} X_d(m) e^{j(2\pi/N)mn}$$

and

$$X_d(m) = \frac{1}{N} \sum_{n=0}^{N-1} x(n) e^{-j(2\pi/N)mn}$$

in Section 2.3.1.

However, it is possible to represent a periodic sequence $\tilde{s}(n)$ by a sum of a finite number of complex exponential sequences of the form $e^{j\omega_k n}$. That is,

$$\tilde{s}(n) = \frac{1}{N} \sum_{k=0}^{N-1} \tilde{S}(k) e^{j(2\pi/N)kn} = \frac{1}{N} \sum_{k=0}^{N-1} \tilde{S}(k) e^{j\omega_k n}$$

holds for all n where ω_k is the kth harmonic frequency obtained as the fundamental frequency $\omega_1 = 2\pi/N$ multiplied by k. The fundamental frequency is found by the following brief consideration. If T_s is the sampling interval to be normalized to unity, then the period T_p of $\tilde{s}(n)$ is

$$T_p = N \cdot T_s = N \quad \text{(s)},$$

so that the fundamental frequency f_1 which is given by the reciprocal of T_p is

$$f_1 = \frac{1}{T_p} = \frac{1}{NT_s} = \frac{1}{N} \quad \text{(Hz)}$$

and f_1 (Hz) can be changed to get the angular frequency by

$$\omega_1 = 2\pi f_1 = \frac{2\pi}{N} \quad (\text{rad s}^{-1}).$$

The equation

$$\tilde{s}(n) = \frac{1}{N} \sum_{k=0}^{N-1} \tilde{S}(k) e^{j\omega_k n} = \frac{1}{N} \sum_{k=0}^{N-1} \tilde{S}(k) e^{j(2\pi/N)kn}$$

is called the discrete Fourier series (DFS) representation of $\tilde{s}(n)$. Here $\tilde{S}(k)$, the coefficient of the kth harmonic component, is expressed using one-period samples of $\tilde{s}(n)$ as

$$\tilde{S}(k) = \sum_{n=0}^{N-1} \tilde{s}(n) e^{-j\omega_k n} = \sum_{n=0}^{N-1} \tilde{s}(n) e^{-j(2\pi/N)kn}.$$

This indicates that $\tilde{S}(k)$ is also a periodic sequence with period N because

$$\tilde{S}(k+lN) = \sum_{n=0}^{N-1} \tilde{s}(n) e^{-j(2\pi/N)(k+lN)n}$$

$$= \sum_{n=0}^{N-1} \tilde{s}(n) e^{-j(2\pi/N)kn} e^{-j2\pi ln} = \tilde{S}(k)$$

holds for any integer l.

To see that $\tilde{S}(k)$ is exactly represented by one period of $\tilde{s}(n)$, we substitute $\tilde{S}(k)$,

$$\tilde{S}(k) = \sum_{n=0}^{N-1} \tilde{s}(n)e^{-j(2\pi/N)kn},$$

into the equation

$$\tilde{s}(n) = \frac{1}{N}\sum_{k=0}^{N-1} \tilde{S}(k)e^{j(2\pi/N)kn}.$$

We have the relation

$$\tilde{s}(n) = \frac{1}{N}\sum_{k=0}^{N-1} \tilde{S}(k)e^{j(2\pi/N)kn} = \frac{1}{N}\sum_{k=0}^{N-1}\sum_{m=0}^{N-1} \tilde{s}(m)e^{-j(2\pi/N)km}e^{j(2\pi/N)kn}$$

$$= \frac{1}{N}\sum_{m=0}^{N-1} \tilde{s}(m)\sum_{k=0}^{N-1} e^{j(2\pi/N)k(n-m)} = \frac{1}{N}\sum_{m=0}^{N-1} \tilde{s}(m)\delta(n-m) = \tilde{s}(n).$$

In a similar manner to the derivation procedure for $\tilde{s}(n)$ from $S(k)$ discussed in Section 5.1, we get

$$\frac{1}{N}\sum_{k=0}^{N-1} \tilde{S}(k)e^{j\omega_k n} = \tilde{s}(n) \qquad \text{for all } n.$$

Thus we can see that the periodic sequence $\tilde{s}(n)$ can be exactly represented by one period of sequence $\tilde{S}(k)$.

Comparing the DFT of a finite-length sequence $s(n)$,

$$S(m) = \sum_{n=0}^{N-1} s(n)e^{-j(2\pi/N)mn} \qquad 0 \le m \le N-1,$$

with the DFS of $\tilde{s}(n)$,

$$\tilde{S}(m) = \sum_{n=0}^{N-1} \tilde{s}(n)e^{-j(2\pi/N)mn} \qquad 0 \le m \le N-1,$$

we can see that the DFT of $s(n)$ is identical to the DFS of $\tilde{s}(n)$ if

$$s(n) = \begin{cases} \tilde{s}(n) & 0 \le n \le N-1 \\ 0 & \text{otherwise.} \end{cases}$$

This indicates that the DFT of a finite-length sequence is the same as the DFS of $\tilde{s}(n)$, assuming that the periodic sequence $\tilde{s}(n)$ is made by concatenating infinite numbers of the finite sequence $s(n)$,

$$\tilde{s}(n) = \sum_{l=-\infty}^{\infty} s(n + lN) \qquad l \text{ integer.}$$

Conversely, if we need to get the finite sequence $s(n)$ from the DFT or DFS, this is realized by extracting one period of the periodic sequence $S(k)$ or $\tilde{S}(k)$. That is,

$$s(n) = \frac{1}{N} \sum_{k=0}^{N-1} S(k)e^{j(2\pi/N)kn} \qquad 0 \leq n \leq N-1,$$

or equivalently,

$$s(n) = \frac{1}{N} \sum_{k=0}^{N-1} \tilde{S}(k)e^{j(2\pi/N)kn} \qquad 0 \leq n \leq N-1.$$

5.3   Properties of DFT

Consider two finite-length time sequences $s_1(n)$ and $s_2(n)$ defined for $0 \leq n \leq N-1$ and their DFTs $S_1(m)$ and $S_2(m)$ defined for $0 \leq m \leq N-1$. Several useful properties of DFTs will be presented in this section. The mathematical operations such as additions and multiplications between two discrete-time Fourier transforms can be obtained because the Fourier transforms are defined in the continuous frequency range from $-\pi$ to $+\pi$. The operations between DFT sequences, however, need a little care, because DFT is of a finite-length sequence and the two DFT sequences to be operated on must be of the same length. In the following we use a symbol $\underset{(N)}{DFT} [\cdot]$ for convenience to represent the N-point discrete Fourier transform of the sequence in the brackets.

5.3.1   Linearity

Since $S_1(m)$ and $S_2(m)$ are defined as

$$S_1(m) = \sum_{n=0}^{N-1} s_1(n)e^{-j(2\pi/N)nm} \qquad 0 \leq m \leq N-1$$

and

$$S_2(m) = \sum_{n=0}^{N-1} s_2(n)e^{-j(2\pi/N)nm} \qquad 0 \leq m \leq N-1,$$

the next relation holds for any complex constants α_1 and α_2:

$$\underset{(N)}{DFT}[\alpha_1 s_1(n) + \alpha_2 s_2(n)] = \alpha_1 S_1(m) + \alpha_2 S_2(m).$$

If the lengths of $s_1(n)$ and $s_2(n)$ are different, say N_1 and N_2, respectively, then DFT must be carried out for the length $N_3 \geq \max(N_1, N_2)$. Moreover, if a particular length $M (\geq N_3)$ of $S_1(m)$ or $S_2(m)$ is required, DFT may be computed for length M after zero padding for $s_1(n)$ by $M - N_1$ zeros and for $s_2(n)$ by $M - N_2$ zeros.

5.3.2 Circular Shift of Sequences

Let $S_d(e^{j\omega})$ be the discrete-time Fourier transform of a finite length-N sequence $s(n)$ with nonzero values in the interval $0 \leq n \leq N - 1$. The discrete-time Fourier transform has the continuous variable ω. The Fourier transform $S_{d,k}(e^{j\omega})$ of the k-point time-shifted $s(n)$ will be

$$S_{d,k}(e^{j\omega}) = \sum_{n=-\infty}^{\infty} s(n-k)e^{-j\omega n} = \sum_{l=-\infty}^{+\infty} s(l)e^{-j\omega(l+k)}$$

$$= e^{-j\omega k} \sum_{l=-\infty}^{+\infty} s(l)e^{-j\omega l} = e^{-j\omega k} S_d(e^{j\omega}),$$

where $l = n - k$. We can see that the discrete-time Fourier transform of a k-point time-shifted sequence equals the discrete-time Fourier transform of the original sequence multiplied by $e^{-j\omega k}$.

In the case of operations using DFT, in order to obtain the k-point time-shifted sequence $s(n - k)$ we use a time-shift operation called 'circular shift' in the interval $0 \leq n \leq N - 1$ which allows the samples going out of one end of the time interval $0 \leq n \leq N - 1$ produced by a shift operation to go back into the opposite end of the interval. An example of a circular shift operation is illustrated in Fig. 5.3.1.

Applying the circular shift operation to the sequence $s(n)$, the DFT $S_k(m)$ of the k-point time-shifted sequence $s(n)$ is

$$S_k(m) = \sum_{n=0}^{N-1} s(n-k)e^{-j(2\pi/N)mn} = \sum_{l=-k}^{N-1-k} s(l)e^{-j(2\pi/N)m(k+l)}$$

$$= e^{-j(2\pi/N)km} \sum_{l=0}^{N-1} s(l)e^{-j(2\pi/N)ml} = e^{-j(2\pi/N)km} S(m),$$

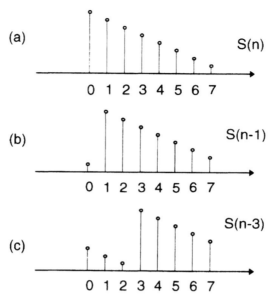

Figure 5.3.1 An example of circular shift operation for $N = 8$: (a) original sequence; (b) and (c) the 1- and 3-point time-shifted sequences, respectively.

where $n - k = l$ and $s(l) = s(l + iN)$ for any integer i because of the circular shift. From the derivation above we see that

$$\underset{(N)}{DFT}[s(n - k)] = e^{-j(2\pi/N)km} \cdot \underset{(N)}{DFT}[S(m)]$$

where the sequence $s(n - k)$ is produced by the circular shift operation.

5.3.3 Multiplication of two DFTs: Circular Convolution Theorem

Let $S_1(m)$ and $S_2(m)$ be the DFTs of two time sequences $s_1(n)$ and $s_2(n)$, respectively. These DFTs and time sequences are of length N. We will find the time sequence $s_3(n)$ whose DFT is the product of $S_1(m)$ and $S_2(m)$. From the definition of DFT, we get

$$S_1(m)S_2(m) = \sum_{n=0}^{N-1} s_1(n)e^{-j(2\pi/N)nm} \cdot [S_2(m)]$$

$$= \sum_{n=0}^{N-1} s_1(n)\{e^{-j(2\pi/N)mn} S_2(m)\}$$

$$= \sum_{n=0}^{N-1} s_1(n) \sum_{k=0}^{N-1} s_2(k - n)e^{-j(2\pi/N)mk}$$

$$= \sum_{k=0}^{N-1} \left\{ \sum_{n=0}^{N-1} s_1(n) s_2(k-n) \right\} e^{-j(2\pi/N)km},$$

or equivalently,

$$S_1(m) S_2(m) = \sum_{k=0}^{N-1} \left\{ \sum_{l=0}^{N-1} s_2(l) s_1(k-l) \right\} e^{-j(2\pi/N)km}.$$

A more detailed derivation will be found in Section 5.5.2.

In the derivation above we used the circular shift for time-shifting, such as $s_1(k-l)$ and $s_2(k-n)$. The equation above shows that the product of the DFTs of two time sequences corresponds to the convolution of the two time sequences, assuming that time shift operations are performed on the circular shift basis discussed in the previous section. Such a convolution is called a *circular convolution*, emphasizing the difference from linear (or ordinary) convolution. Then we have

$$\underset{(N)}{DFT}[s_1(n) \otimes s_2(n)] = \underset{(N)}{DFT}[s_1(n)] \cdot \underset{(N)}{DFT}[s_2(n)]$$

where we have used the symbol $\otimes$ to indicate circular convolution. This result is called the *circular convolution theorem* which is often referred to in calculating time-invariant linear system responses in the frequency domain.

Example 5.3.1

An example of a circular convolution in the interval $0 \le n \le 4$ will be shown for two sequences as

$$s_1(n) = 1 \qquad 0 \le n \le 4$$

and

$$s_2(n) = \delta(n) \qquad 0 \le n \le 4.$$

The 5-point circular convolution for $s_1(n)$ and $s_2(n)$ becomes

$$s_3(l) = \sum_{n=0}^{4} s_1(n) s_2(l-n) = \sum_{n=0}^{4} s_2(l-n) = \sum_{n=0}^{4} \delta(l-n) = 1,$$

where $0 \le l \le 4$. The DFTs of $s_1(n)$ and $s_2(n)$ are

$$S_1(m) = \sum_{n=0}^{4} s_1(n) e^{-j(2\pi/5)nm}$$

$$= s_1(0) + s_1(1)e^{-j(2\pi/5)m} + s_1(2)e^{-j(2\pi/5)2m}$$
$$+ s_1(3)e^{-j(2\pi/5)3m} + s_1(4)e^{-j(2\pi/5)4m}$$
$$= 1 + e^{-j(2\pi/5)m} + e^{-j(2\pi/5)2m} + e^{-j(2\pi/5)3m} + e^{-j(2\pi/5)4m}$$
$$= \frac{1 - e^{-j(2\pi/5)5m}}{1 - e^{-j(2\pi/5)m}} = 0 \qquad \text{for } 0 < m \le 4.$$

For $m = 0$ we have

$$S_1(0) = \sum_{n=0}^{4} s_1(n) = 5.$$

Thus, we get

$$S_1(m) = \sum_{n=0}^{4} s_1(n)e^{-j(2\pi/5)nm} = 5\delta(m) \qquad \text{for } 0 \le m \le 4.$$

Similarly,

$$S_2(m) = \sum_{n=0}^{4} s_2(n)e^{-j(2\pi/5)nm} = s_2(0) = 1 \qquad 0 \le m \le 4.$$

Let $S_3(m)$ be the DFT of $s_3(l)$ and we have

$$S_3(m) = \sum_{l=0}^{4} s_3(l)e^{-j(2\pi/5)lm} = 5\delta(m) \qquad 0 \le m \le 4.$$

Therefore, the relation

$$S_3(m) = S_1(m) \cdot S_2(m)$$

holds. The procedure is illustrated in Fig. 5.3.2.

5.3.4   DFT of the Product of Two Sequences

Let two real sequences be $s_1(n)$ and $s_2(n)$ both defined in the interval $0 \le n \le N - 1$ and the corresponding DFTs be $S_1(m)$ and $S_2(m)$ also defined in $0 \le m \le N - 1$. Consider the DFT of the product of $s_1(n)$ and $s_2(n)$,

$$\underset{(N)}{DFT}[s_1(n)s_2(n)]$$

$$= \sum_{n=0}^{N-1} s_1(n)s_2(n)e^{-j(2\pi/N)nm}$$

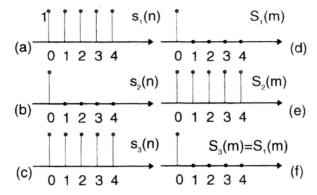

Figure 5.3.2 An example of the circular convolution theorem. (a), (b), (c): $s_1(n)$, $s_2(n)$, and $s_3(n) = s_1(n) \times s_2(n)$. (d), (e), (f): $S_1(m)$, $S_2(m)$ and $S_3(m) = S_1(m)S_2(m)$. The heights of all pulses are 1 except for (d) and (f) with height 5.

$$= \sum_{n=0}^{N-1} \frac{1}{N^2} \left\{ \sum_{l=0}^{N-1} S_1(l)e^{j(2\pi/N)ln} \sum_{k=0}^{N-1} S_2(k)e^{j(2\pi/N)kn} \right\} e^{-j(2\pi/N)nm}$$

$$= \sum_{n=0}^{N-1} \frac{1}{N^2} \left\{ \sum_{l=0}^{N-1} S_1(l) \sum_{k=0}^{N-1} S_2(k)e^{j(2\pi/N)(k+l)n} \right\} e^{-j(2\pi/N)nm}$$

$$= \sum_{n=0}^{N-1} \frac{1}{N^2} \left\{ \sum_{l=0}^{N-1} S_1(l) \sum_{p=l}^{N-1+l} S_2(p-l)e^{j(2\pi/N)pn} \right\} e^{-j(2\pi/N)nm}$$

$$= \sum_{n=0}^{N-1} \frac{1}{N^2} \left\{ \sum_{l=0}^{N-1} S_1(l) \sum_{p=0}^{N-1} S_2(p-l)e^{j(2\pi/N)pn} \right\} e^{-j(2\pi/N)nm}$$

$$= \frac{1}{N^2} \sum_{n=0}^{N-1} \left\{ \sum_{p=0}^{N-1} \sum_{l=0}^{N-1} S_1(l)S_2(p-l)e^{j(2\pi/N)pn} \right\} e^{-j(2\pi/N)nm}$$

$$= \frac{1}{N} \sum_{l=0}^{N-1} S_1(l) \left\{ \frac{1}{N} \sum_{p=0}^{N-1} \sum_{n=0}^{N-1} S_2(p-l)e^{j(2\pi/N)(p-m)n} \right\}$$

$$= \frac{1}{N} \sum_{l=0}^{N-1} S_1(l) \left[\sum_{p=0}^{N-1} S_2(p-l) \left\{ \frac{1}{N} \sum_{n=0}^{N-1} e^{j(2\pi/N)(p-m)n} \right\} \right]$$

$$= \frac{1}{N} \sum_{l=0}^{N-1} S_1(l) \left[\sum_{p=0}^{N-1} S_2(p-l)\delta(p-m) \right]$$

$$= \frac{1}{N} \sum_{l=0}^{N-1} S_1(l) S_2(m - l).$$

In this derivation process circular shift operations

$$\sum_{p=l}^{N-1+l} S_2(p - l) e^{j(2\pi/N)pn} = \sum_{p=0}^{N-1} S_2(p - l) e^{j(2\pi/N)pn}$$

in the frequency domain and the orthogonality condition for the complex exponential functions

$$\frac{1}{N} \sum_{n=0}^{N-1} e^{j(2\pi/N)(p-m)n} = \delta(p - m)$$

is applied. From the result we can see that the DFT of a product of two N-length time sequences is obtained by the circular convolution of two frequency-domain sequences corresponding to the two time sequences divided by N.

5.3.5   Parseval's Equation

From the relation obtained in Section 5.3.4,

$$\sum_{n=0}^{N-1} s_1(n) s_2(n) e^{-j(2\pi/N)nm} = \frac{1}{N} \sum_{l=0}^{N-1} S_1(l) S_2(m - l),$$

we get

$$\sum_{n=0}^{N-1} s^2(n) = \frac{1}{N} \sum_{l=0}^{N-1} S(l) S(-l)$$

assuming that $m = 0$, $s_1(n) = s_2(n) = s(n)$, and $S_1(l) = S_2(l) = S(l)$. Since $S(-l)$ is complex and its imaginary part is an odd function, we have

$$S^*(-l) = S(l),$$

and we can obtain the next relation,

$$\sum_{n=0}^{N-1} s^2(n) = \frac{1}{N} \sum_{l=0}^{N-1} |S(l)|^2.$$

This is called Parseval's equation, which is an equivalent equation derived in the z-domain in Section 3.7.

5.4   Fast Fourier Transform (FFT) Algorithm for DFT Computation

An N-point DFT was shown in Section 5.1 as

$$X(m) = \sum_{n=0}^{N-1} x(n)e^{-j(2\pi/N)mn}$$

where $x(n)$ is a sequence of numbers and $X(m)$ is a sequence of the corresponding DFT, both of which have a length N. In order to obtain $X(m)$ for all m, it is clear from the above definition that we need N^2 multiplications, provided that the constants

$$e^{-j(2\pi/N)k} \qquad \text{for } 0 \leq k \leq N-1$$

are all obtained in advance of the DFT calculation.

Efficient algorithms for computing DFT called fast Fourier transform (FFT) algorithms have been developed and widely used in implementing both software and hardware. According to the FFT algorithm we get an N-point DFT by $N \log_2 N$ multiplications if N is a power of 2. Many books present these algorithms in great detail. Accordingly, in this section we present only a brief description of the basic idea of the algorithms for the fast Fourier transform.

There are two methods for FFT which are based on a decimation in time and on a decimation in frequency. The following discussion is limited to a decimation in time method. To simplify the notation we write

$$W_N^m = e^{-j(2\pi/N)m} \qquad 0 \leq m \leq N-1.$$

In the decimation in time method we divide the right-hand side of the equation

$$X(m) = \sum_{n=0}^{N-1} x(n)W_N^{nm}$$

into two parts as

$$X(m) = \sum_{n=0}^{N-1} x(n)e^{-j(2\pi/N)mn} = \sum_{n=0}^{N-1} x(n)W_N^{mn}$$

$$= \sum_{n=0}^{(N/2)-1} x(2n)W_N^{m \cdot 2n} + \sum_{n=0}^{(N/2)-1} x(2n+1)W_N^{m(2n+1)}$$

$$\text{for } 0 \leq m \leq N-1.$$

We can rewrite $X(m)$ as

$$X(m) = \sum_{n=0}^{(N/2)-1} x_1(n)W_{N/2}^{mn} + W_N^m \sum_{n=0}^{(N/2)-1} x_2(n)W_{N/2}^{mn}$$

$$0 \leq m \leq N-1,$$

where we set

$$\left.\begin{array}{l} x_1(n) = x(2n) \\ x_2(n) = x(2n+1) \end{array}\right\} \quad \text{for } 0 \leq n \leq \frac{N}{2} - 1$$

and the property

$$W_N^2 = (e^{-j(2\pi/N)})^2 = e^{-j2\pi/(N/2)} = W_{N/2}.$$

is used. By setting

$$P(m) = \sum_{n=0}^{(N/2)-1} x_1(n)W_{N/2}^{mn} \quad 0 \leq m \leq N-1$$

and

$$Q(m) = \sum_{n=0}^{(N/2)-1} x_2(n)W_{N/2}^{mn} \quad 0 \leq m \leq N-1,$$

we obtain

$$X(m) = P(m) + W_N^m Q(m) \quad 0 \leq m \leq N-1.$$

If $P(m)$ and $Q(m)$ are already given for $0 \leq m \leq N-1$, we can obtain $X(m)$ for all m ($0 \leq m \leq N-1$) by N additions and N multiplications. $P(m)$ and $Q(m)$ are $(N/2)$-point DFTs for the even-numbered and odd-numbered sequences of the original sequence $x(n)$, respectively. The procedure to get $X(m)$ from $P(m)$ and $Q(m)$ shown above is illustrated in Fig. 5.4.1.

In a similar fashion to the procedure for calculating $X(m)$ ($0 \leq m \leq N-1$) by the combination of $(N/2)$-point DFTs, we can compute $P(m)$ by combination of $(N/4)$-point DFTs of the sequences $x_{11}(n)$ and $x_{12}(n)$, which are composed of the even- and odd-numbered samples of $x_1(n)$, respectively. The sequences are

$$x_{11}(n) = x_1(2n) = x(4n) \quad 0 \leq n \leq \frac{N}{4} - 1$$

and

$$x_{12}(n) = x_1(2n+1) = x(4n+2) \quad 0 \leq n \leq \frac{N}{4} - 1.$$

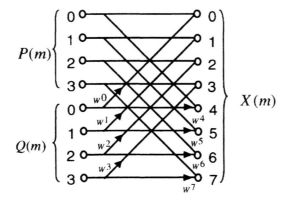

Figure 5.4.1 The procedure to obtain $X(m)$ from $P(m)$ and $Q(m)$ for $m = 8$ where signal flow left to right indicates addition and multiplication.

We can compute $P(m)$ for $0 \leq m \leq (N/2) - 1$ as

$$P(m) = P_1(m) + W_{N/2}Q_1(m) \qquad 0 \leq m \leq \frac{N}{2} - 1,$$

where

$$P_1(m) = \sum_{n=0}^{(N/4)-1} x_{11} W_{N/4}^{mn} \qquad 0 \leq m \leq \frac{N}{4} - 1$$

and

$$Q_1(m) = \sum_{n=0}^{(N/4)-1} x_{12} W_{N/4}^{mn} \qquad 0 \leq m \leq \frac{N}{4} - 1.$$

$Q(m)$ is similarly computed from the DFT of the two sequences of length $N/4$,

$$x_{21}(n) = x_2(2n) = x(4n + 1) \qquad 0 \leq n \leq \frac{N}{4} - 1$$

and

$$x_{22}(n) = x_2(2n + 1) = x(4n + 3) \qquad 0 \leq n \leq \frac{N}{4} - 1.$$

In this procedure to get $P(m)$ for $0 \leq m \leq (N/2) - 1$, $N/2$ multiplications are required provided that $P_1(m)$ and $Q_1(m)$ are given. Since the same number of multiplications as $P(m)$ are needed to obtain $Q(m)$ for $0 \leq m \leq (N/2) - 1$, the total number of multiplications to get $P(m)$ and $Q(m)$ is $(N/2) \times 2 = N$, which is the same number as when we compute

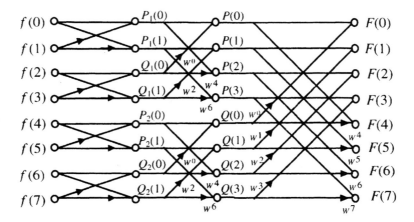

Figure 5.4.2 The complete procedure of the decimation-in-time FFT for $N = 2^3 = 8$ and $W_8 = e^{-j(2\pi/8)}$.

$X(m)$ for $0 \le m \le N - 1$. The total procedure to get $X(m)$ is illustrated in Fig. 5.4.2 for $N = 2^3 = 8$.

The FFT algorithm based on decimation in time repeats the procedure in which an N-point DFT is calculated by the combination of two $(N/2)$-point DFTs. If N is a power of 2, for example $N = 2^R$, then after R repetitions of the procedure, one-point DFT is reached. The one-point DFT is at the sample point itself. At the rth repetition step $(N/2^r)$-point DFT is carried out 2^r times to form N-point DFT. Thus the DFT of the $N = 2^R$ point sequence can be obtained by $NR = N \log_2 N$ multiplications.

Further efficiency can be obtained by a small change of the procedure above. Since $P(m)$ and $Q(m)$ are periodic with the period $N/2$, i.e.,

$$P\left(m + \frac{N}{2}\right) = \sum_{n=0}^{(N/2)-1} x_1(n)W_{N/2}^{[m+(N/2)]n}$$

$$= \sum_{n=0}^{(N/2)-1} x_1(n)W_{N/2}^{mn} \cdot W_{N/2}^{Nn/2}$$

$$= \sum_{n=0}^{(N/2)-1} x_1(n)W_{N/2}^{mn} = P(m)$$

and similarly,

$$Q\left(m + \frac{N}{2}\right) = Q(m),$$

where

$$W_{N/2}^{Nn/2} = (e^{-j2\pi/(N/2)})^{Nn/2} = e^{-j2n\pi} = 1,$$

we get

$$X(m) = P(m) + W_N^m Q(m) \qquad 0 \le m \le \frac{N}{2} - 1$$

and

$$X\left(m + \frac{N}{2}\right) = P\left(m + \frac{N}{2}\right) + W_N^{m+(N/2)} Q\left(m + \frac{N}{2}\right)$$

$$= P(m) - W_N^m Q(m) \qquad \text{for } 0 \le m \le \frac{N}{2} - 1,$$

where the relation

$$W_N^{m+(N/2)} = e^{-j(2\pi/N)[m+(N/2)]} = e^{-j(2\pi/N)m} e^{-j\pi} = -W_N^m$$

is used for $0 \le m \le N/2$. The term $W_N^m Q(m)$ can be calculated for each m $(0 \le m \le N/2 - 1)$ and used for obtaining both the first and the second half of $X(m)$. Therefore $N/2$ multiplications are required including the special cases such as

$$W_N = \pm 1 \qquad \text{or} \qquad \pm j.$$

In this algorithm, $N/2$ multiplications are required to get N-point $X(m)$ at each stage. So the total multiplication number reduces to $\frac{1}{2}RN$.

5.5 Convolution using DFT

In Chapters 2 and 3 we observed that a convolution sum of two time sequences is represented by a product of the two corresponding discrete-time transforms or z-transforms. In this section we will find a relation between two sequences and their DFTs.

5.5.1 Linear Convolution

Consider two time-sequences $s_1(n)$ and $s_2(n)$ defined over the interval $-\infty < n < +\infty$. We have a convolution sum $s_3(n)$,

$$s_3(n) = \sum_{l=-\infty}^{\infty} s_1(l)s_2(n-l) = \sum_{l=-\infty}^{\infty} s_2(l)s_1(n-l).$$

If we perform a time shift operation in an ordinary sense, this operation is referred to as a linear convolution. If $s_1(n)$ and $s_2(n)$ are finite sequences as

$$s_1(n) = \begin{cases} \text{nonzero} & 0 \le n \le N_1 - 1 \\ 0 & \text{otherwise} \end{cases}$$

and

$$s_2(n) = \begin{cases} \text{nonzero} & 0 \le n \le N_2 - 1 \\ 0 & \text{otherwise,} \end{cases}$$

then

$$s_3(n) = \sum_{l=0}^{N_1-1} s_1(l)s_2(n-l) = \sum_{l=0}^{N_2-1} s_2(l)s_1(n-l).$$

The total number of the nonzero part of $s_3(n)$ will be

$$N_3 = N_1 + N_2 - 1.$$

Example 5.5.1

The sequences $s_1(n)$ and $s_2(n)$ are of length 3 and 5, respectively. The linear convolution $s_3(n)$ is

$$s_3(n) = \sum_{l=0}^{2} s_1(l)s_2(n-l) = \sum_{l=0}^{4} s_2(l)s_1(n-l).$$

Figure 5.5.1 illustrates the result. We can see that the length of $s_3(n)$ is $7(= 3 + 5 - 1)$.

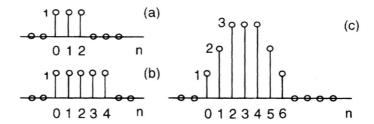

Figure 5.5.1   An example of a linear convolution: (a) $s_1(n)$; (b) $s_2(n)$; (c) $s_3(n) = s_1(n) * s_2(n)$.

5.5.2   Circular Convolution

Consider two sequences $s_1(n)$ and $s_2(n)$ defined in the interval $0 \le n \le N - 1$. The circular convolution of the two sequences is defined as

$$s_3(l) = \sum_{n=0}^{N-1} s_1(n)s_2(l-n) = \sum_{n=0}^{N-1} s_2(n)s_1(l-n),$$

where $s_3(l)$ is the resultant sequence of the convolution and the time shift operation for $s_1(n)$ and $s_2(n)$ is done on the circular shift basis as

$$s_1(n \pm N) = s_1(n)$$

and

$$s_2(n \pm N) = s_2(n).$$

Then we have the circular convolution theorem as discussed in Section 5.3.3,

$$\underset{(N)}{DFT}[s_3(n)] = \underset{(N)}{DFT}[s_1(n)] \cdot \underset{(N)}{DFT}[s_2(n)].$$

Let the DFTs of $s_1(n)$, $s_2(n)$ and $s_3(n)$ be $S_1(m)$, $S_2(m)$ and $S_3(m)$, respectively. Then the equation above is written as

$$S_3(m) = S_1(m) \cdot S_2(m) \qquad 0 \leq m \leq N - 1.$$

The multiplications should be performed term by term. The proof of the theorem is as follows. Taking the DFT of the sequence $s_3(n)$ obtained by the circular convolution of $s_1(n)$ and $s_2(n)$, we get

$$S_3(m) = \sum_{n=0}^{N-1} s_3(n)e^{-j(2\pi/N)nm}$$

$$= \sum_{n=0}^{N-1} \left\{ \sum_{l=0}^{N-1} s_1(l)s_2(n-l) \right\} e^{-j(2\pi/N)nm}.$$

Interchanging the order of the two summations,

$$S_3(m) = \sum_{l=0}^{N-1} s_1(l) \sum_{n=0}^{N-1} s_2(n-l)e^{-j(2\pi/N)nm}.$$

Using the circular shift property with interval N discussed in the Section 5.3.2, we get

$$S_3(m) = \sum_{l=0}^{N-1} s_1(l)e^{-j(2\pi/N)lm} \cdot \sum_{n=0}^{N-1} s_2(n)e^{-j(2\pi/N)nm} = S_1(m)S_2(m).$$

The equation above shows the result we want. Circular convolution is similar to the linear convolution in the sense that one of two sequences is time-reversed, time-shifted, and multiplied by other sequences term by term, and the products are summed. But it is different in the sense that the time reversal and time shift operations are done circularly in the interval $0 \leq n \leq N - 1$. This is due to the fact that DFT of a finite-length sequence is equivalent to DFS of a periodic sequence one period of which is identical to the original finite sequence.

Example 5.5.2

Let two sequences be

$$s_1(n) = \begin{cases} 1 & 0 \leq n \leq 3 \\ 0 & \text{otherwise} \end{cases}$$

and

$$s_2(n) = \begin{cases} 1 & 0 \leq n \leq 5 \\ 0 & \text{otherwise.} \end{cases}$$

The circular convolution follows from the definition as

$$s_3(l) = \sum_{n=0}^{N-1} s_1(n)s_2(l - n),$$

where N is the interval within which the circular shift is done. Figure 5.5.2 shows the results of convolution for several interval length N. We can see the circular convolution results for $N = 5, 6, 7$ are all different and in the case of $N = 7$ the circular convolution is identical to the linear convolution. Although the circular convolution results for $N > 7$ are not shown in Fig. 5.5.2, these results are all identical to the linear convolution.

We will show the reason why a linear convolution can be obtained by calculating a circular convolution under the condition that the time shift of the sequences is carried out circularly within the interval $0 \leq n \leq N - 1$ ($N \geq N_1 + N_2 - 1$), where N_1 and N_2 are the lengths of two sequences to be convolved with each other. Figure 5.5.3 shows the circular convolution procedure for two sequences $s_1(n)$ and $s_2(n)$ defined by

$$s_1(n) = \begin{cases} 1 & 0 \leq n \leq 2 \\ 0 & \text{otherwise} \end{cases}$$

and

$$s_2(n) = \begin{cases} 1 & 0 \leq n \leq 4 \\ 0 & \text{otherwise.} \end{cases}$$

In Fig. 5.5.3 we choose $N = 7$ ($= 3 + 5 - 1$) and calculate

$$s_3(l) = \sum_{n=0}^{N-1} s_1(n)s_2(l - n) = \sum_{n=0}^{6} s_1(n)s_2(l - n).$$

The samples shown as dotted pulses show the time-reversed sequence $s_2(l - n)$ and these pulses are circularly shifted to the opposite end of the

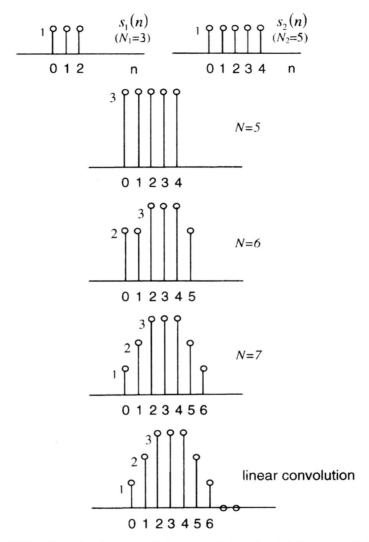

Figure 5.5.2 The circular convolution of $s_1(n)$ and $s_2(n)$ for several different length of intervals where $N = 5, 6, 7, 8$ are illustrated together with the linear-convolution result.

interval $0 \le n \le N - 1$. For example, when $l = 0$, the samples of $s_2(-n)$ for $n = 4 \sim 1$ are circularly shifted to $s_2(n)$ for $n = 3 \sim 6$, respectively. Since $s_1(n) = 0$ for $n = 3 \sim 6$, the partial sum of products to be obtained in the summation above for $n = 3 \sim 6$ is zero. Therefore, the part of sequence $s_2(n)$ circularly shifted to $n = 3 \sim 6$ makes no contribution to the convolution result $s_3(0)$. For l other than 0 a similar situation can be seen in Fig. 5.5.3. By choosing $N \ge N_1 + N_2 - 1$ we can avoid the effect

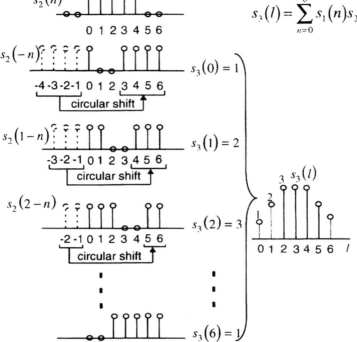

$$s_3(l) = \sum_{n=0}^{6} s_1(n)s_2(l-n)$$

circular convolution
$N = 7$

Figure 5.5.3   The circular convolution of $s_1(n)$ and $s_2(n)$ with the circular shift interval 7.

of circular shift on the resultant convolution and consequently we can get the linear convolution.

5.5.3   Linear Convolution of Two Time Sequences Based on DFT

We have two ways of calculating the linear convolution. One is performed in the time domain; the other is done in the frequency domain. Consider two finite-length time sequences $s_1(n)$ and $s_2(n)$ as

$$s_1(n) = \begin{cases} \text{nonzero} & 0 \leq n \leq N_1 - 1 \\ 0 & \text{otherwise} \end{cases}$$

and

$$s_2(n) = \begin{cases} \text{nonzero} & 0 \leq n \leq N_2 - 1 \\ 0 & \text{otherwise.} \end{cases}$$

In the time-domain method we calculate directly the next form

$$y(n) = s_1(n) * s_2(n) = \sum_{l=-\infty}^{\infty} s_1(l)s_2(n-l)$$

$$= \sum_{l=0}^{N_1-1} s_1(l)s_2(n-l) = \sum_{l=0}^{N_2-1} s_2(l)s_1(n-l).$$

The convolution sequence $y(n)$ can be seen as the output of a linear time-invariant system with the impulse response $s_1(n)$ when the input sequence $s_2(n)$ is applied. Therefore, the resultant sequence $y(n)$ will have the length $(N_1 + N_2 - 1)$ and the direct calculation of $y(n)$ shown above requires $N_1 N_2$ multiplications.

On the other hand, in the frequency-domain method the two time sequences are transformed to DFTs and multiplied term by term. Then the DFT obtained is again transformed back into the time domain using inverse DFT. As described in Section 5.4, we have an efficient calculation algorithm for the DFT called FFT (fast Fourier transform). Using an FFT we can achieve a substantial reduction of computing time of a DFT. The time consumed by the computation of an N-point DFT is approximately proportional to N^2, while using FFT it is reduced to approximately $N \log_2 N$ under the condition that N is a power of 2.

The first step of the frequency-domain method is to determine the length of the DFT. The sequences $s_1(n)$ and $s_2(n)$ have, in general, different lengths N_1 and N_2, so we give them the same length N by zero-padding where $N \geq N_1$ and $N \geq N_2$. Since $y(n)$ is of length $N_1 + N_2 - 1$, as stated above, N must satisfy the inequality

$$N \geq N_1 + N_2 - 1.$$

We choose N as the smallest power of 2 satisfying this inequality in order to use the FFT algorithm. For example when $N_1 = 15$ and $N_2 = 20$, N is chosen as $32 = 2^5$.

The next step is to perform DFTs of the sequences and to multiply them. Thus we get the product of the two DFTs as

$$S_1(k)S_2(k) \qquad 0 \leq k \leq N-1,$$

where $S_1(k)$ and $S_2(k)$ are the DFTs of $s_1(n)$ and $s_2(n)$, respectively.

The final step is to get the inverse DFT (denoted by IDFT); we get

$$y(n) = \underset{(N)}{IDFT}\{S_1(k)S_2(k)\} \qquad 0 \leq n \leq N-1.$$

Here it should be noted that, as discussed in Section 5.3.3, a product of two DFTs corresponds to the circular convolution of the two time-sequences. That is, the sequence $y(n)$ obtained above is not the linear convolution but the circular convolution of $s_1(n)$ and $s_2(n)$. There is no trouble, however, getting the linear convolution, because we use N larger than the length $(N_1 + N_2 - 1)$ of $y(n)$ obtained by the linear convolution.

The frequency-domain method needs three DFT operations. If $s_1(n)$ and $s_2(n)$ have length of N which is a power of 2 satisfying $N \geq N_1 + N_2 - 1$, where N_1 and N_2 are the lengths of the sequences $s_1(n)$ and $s_2(n)$, respectively, we can use the FFT algorithm for the DFT calculations. Consequently, the total number of multiplications is approximately proportional to $3N \log_2 N$. For example, when $N_1 = 500$ and $N_2 = 100$, we can choose $N = 1024$ and see that the number of multiplications required in the frequency-domain method for the convolution is less than that for the time-domain method as follows:

$Time\text{-}domain$: $N_1 N_2 = 50\,000$ multiplications

$Frequency\text{-}domain$: $3N \log_2 N = 30\,720$ multiplications.

The frequency-domain method is more efficient than the time-domain method for calculating convolutions.

5.6   Linear System Response Obtained by Convolution

An output sequence of a linear time-invariant system is obtained by the convolution of the impulse response sequence $h(n)$ of the system and the input sequence $x(n)$. Here we deal with FIR (finite impulse response) systems. That is, we have the form

$$y(n) = h(n) * x(n)$$

where $h(n)$ is of finite-length and $x(n)$ is indefinitely long. We often encounter this situation in applications such as filtering speech signals through a low-pass filter. Even in that situation we want to use the frequency-domain method to achieve efficient calculations of the convolution.

The length of the input sequence $x(n)$ is, however, indefinitely long, so it is neither possible nor practical to store whole the input sequence data in a computer memory in order to perform DFTs. Therefore, we have to divide the input sequence into finite-length segments to calculate the convolution between each segment and $h(n)$, and we have to connect the convolution results without any distortions at the connection points between segments. We have two practical methods, depending on the difference of the method of connection.

5.6.1 Overlap-add Method

The procedure of the overlap-add is illustrated in Fig. 5.6.1. Suppose that the sequence $h(n)$ is an N_1-length impulse response sequence of a linear system and $x(n)$ is the input sequence. The sequence $x(n)$ is divided into segments of length N_2. We assume $N_2 > N_1$. The first step is to calculate the linear convolution as

$$y_1(n) = h(n) * x_1(n)$$

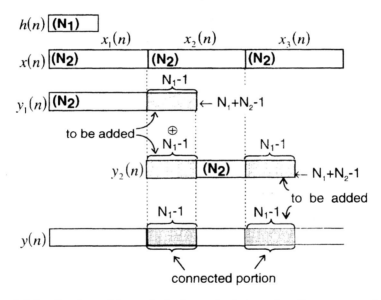

Figure 5.6.1 Overlap-add method for calculating the linear convolution $y_1(n) = h(n) * x_1(n)$.

where $x_1(n)$ is the first segment of $x(n)$ and $y_1(n)$ is the resultant sequence. The first N_2 samples of $y_1(n)$ are the correct result, but the last $N_1 - 1$ samples are incomplete, since the $N_1 - 1$ samples are the output sequence when the input is 0.

The second step is to calculate

$$y_2(n) = h(n) * x_2(n)$$

for the second segment of $x_2(n)$. The output sequence $y_2(n)$ is of $N_1 + N_2 - 1$ length and the top $N_1 - 1$ samples of $y_2(n)$ are obtained at the zero (without memory of the preceding segment) state of the system. So, according to the superposition principle, we can get the correct convolution by adding the $N_1 - 1$ tail samples of $y_1(n)$ and $N_1 - 1$ head samples of $y_2(n)$ (the third step). In this way we can calculate the correct output samples at the connection part of the $y_1(n)$ and $y_2(n)$.

By using a formal expression we have the output sequence $y(n)$ as

$$y(n) = \sum_{i=1} y_i(n - (i - 1)N_2),$$

where i is the segment number of the input sequence and $y_i(n)$ is the ith output segment. By repeating the three steps mentioned above we can get the linear convolution of $h(n)$ and $x(n)$. This method does not necessarily require the frequency-domain convolutions. The term 'overlap-add method' comes from the fact that the overlapped parts of $y_1(n)$ and $y_2(n)$ are connected without any distortion.

5.6.2 Overlap-save Method

The procedure of this method is illustrated in Fig. 5.6.2. Consider the linear convolution for an impulse response $h(n)$ $(0 \le n \le N_1 - 1)$ with an input $x(n)$. The input sequence $x(n)$ is divided into short segments of length N_2, where $N_2 > 2N_1$. These segments are numbered such that

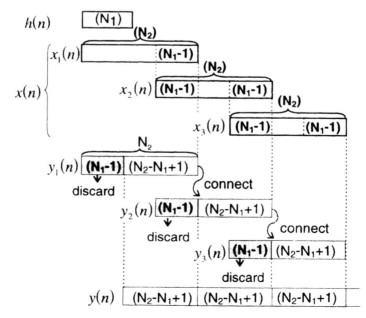

Figure 5.6.2   Overlap-save method where $N_2 > 2N_1$ is assumed.

$x_i(n) \, (i = 1, 2, \ldots)$, and adjacent pairs of segments are overlapped by $N_1 - 1$ samples as

$$x_i(n : N_2 - N_1 + 1 \leq n \leq N_2 - 1) = x_{i+1}(n : 0 \leq n \leq N_1 - 2).$$

Here we assume that the time origin of every input segment is at the first sample. This is also assumed for every output segment $y_i(n)$.

The first step of this method is to perform the circular convolution of $h(n)$ and $x_i(n)$,

$$y_i(n) = h(n) \otimes x_i(n) = \sum_{k=0}^{N_2-1} x_i(k) h(n - k),$$

where $y_i(n)$ is the ith output sequence obtained by this convolution and the time shift of $h(k - n)$ is done on a circular shift basis with the interval $0 \leq k \leq N_2 - 1$. As mentioned in Section 5.3.3, the convolution obtained through frequency-domain processing based on the convolution theorem is the circular convolution.

This property is used efficiently in this overlap-save method. The top $(N_1 - 1)$-length samples of $y_i(n)$ obtained by the circular convolution are not equal to the top $(N_1 - 1)$-length sequence obtained by the linear convolution. This is because we use circular shift operations for time shifting. So the samples $y_i(n) \, (0 \leq n \leq N_1 - 2)$ are discarded.

The next step is to calculate

$$y_{i+1}(n) = h(n) \otimes x_{i+1}(n).$$

Then the top $N_1 - 1$ samples of $y_{i+1}(n) \, (0 \leq n \leq N_1 - 2)$ are also discarded. Finally, we connect $y_{i+1}(n)$ as illustrated in Fig. 5.5.2. The last $N_1 - 1$ samples of the ith output sequence $y_i(n)$ give the correct linear convolution result, so that we can get the linear convolution for the input segments $x_i(n)$ and $x_{i+1}(n)$ by connecting

$$y_i(n) \qquad \text{for } N_2 - N_1 + 1 \leq n \leq N_2 - 1$$

and

$$y_{i+1}(n) \qquad \text{for } N_2 - N_1 + 1 \leq n \leq N_2 - 1.$$

By repeating the three steps we can get the linear convolution of the input sequence $x(n)$ with indefinite length.

5.7 Summary

In this chapter we derived the discrete Fourier transform (DFT) of sequences from the discrete-time Fourier transform. We showed that the

DFT is obtained by sampling the discrete-time Fourier transform at equally spaced frequency intervals and that the DFT of a finite-length sequence is identical to the sequence of discrete Fourier series coefficients of a periodic sequence made by combining copies of the finite-length sequence. We also showed the properties of the DFT, including the circular shift and the circular convolution properties. An efficient algorithm for computing DFTs called the fast Fourier transform (FFT) algorithm was introduced. We showed that the computing time required for linear convolutions in the frequency domain is less than that necessary for the direct computation of the linear convolution in the time domain. Finally, we discussed practical methods for computing linear convolutions where an indefinitely long sequence is convolved with a finite-length sequence.

6 Transfer Function Models and Wave Equations

6.0 Introduction

This chapter summarizes the fundamentals of acoustic wave theory which constructs an important basis for acoustic signal processing. Acoustic signal processing is performed based on discrete system models; however, acoustic signals are produced by acoustic and mechanical processes of continuous media which are governed by wave equations. In this chapter we describe the wave equations and acoustic transfer functions (TFs). Although a wave equation may not seem to be related to acoustic signal processing technology, sound field control in a three-dimensional (3-D) space requires sound wave radiation and propagation control.

The sound wave radiation and propagation phenomena are described by the wave equation. We will derive the plane wave and general wave equations, sound wave reflection, simple models of sources, sound power output, driving point impedance, and image theory. We will summarize eigenfrequencies and eigenfunctions, Hermitian operators and orthogonality, Green's functions, and Kirchhoff's integral solution of the Helmholtz equation using Green's functions. Kirchhoff's integral formula using Green's function is a classical result; however, it will provide a fundamental scheme for multichannel sound field control.

We will use Green's function as a model of transfer functions that represents the deep structure of acoustic systems. This is because Green's function can be interpreted as Fourier transform of the impulse response of an acoustic system which is excited by a point source. Models such as 1-D acoustic pipes and vibrations of 2-D circular membranes give a useful insight into the deep structures of acoustic systems when we construct appropriate models using discrete observation data.

Readers who are interested in the detailed wave theory are invited to read accepted textbooks such as *Hydrodynamics* (Sir H. Lamb, Dover Publications, New York, 1945), *Theory of Sound* (J.W.S. Rayleigh, Dover Publications, 1945), *Theoretical Acoustics* (P.M. Morse and

K.U. Ingard, Princeton University Press, New Jersey, 1968) or *Principals of Vibration and Sound* (T. Rossing and H. Fletcher, Springer Verlag, New York, 1995. Wave solution analyses in Sections 6.1.3, 6.1.6, and 6.5.9 in particular are described following Lamb's *Hydrodynamics*.

6.1   Wave Equations

6.1.1   State Equation and Elasticity of an Ideal Gas

The state of gas can be described using three variables known as pressure p (Pa), volume density ρ (kg m^{-3}), and temperature θ (K). If we take the mass of a gas M (kg) and the volume V (m^3) occupied by the gas, we can define the specific volume v (m^3 kg^{-1}) as

$$\rho = \frac{M}{V} = \frac{1}{v} \quad (\text{kg m}^{-3}),$$

where

$$v = \frac{V}{M} \quad (\text{m}^3 \text{kg}^{-1}).$$

In the case of an ideal gas, we have the relationship

$$pv = R\theta,$$

where R is the gas constant depending on the gas. This relationship is called the *state equation* of an ideal gas. In an isothermal process where the temperature is constant, we have

$$pv = \text{constant}.$$

In the adiabatic case where there is no thermal energy exchange between the system of interest and its environment, we get another relationship. Let us take the thermal energy $P\,\delta p$ (J) which is required for a δp pressure increase when the gas has unit mass and is maintained at constant specific volume. P is an unknown constant independent of the pressure variation. Taking the specific heat c_v under constant specific volume, we get

$$P\,\delta p = c_v\,\delta\theta,$$

where $\delta\theta$ denotes the increase of temperature of the gas due to the thermal energy $P\,\delta p$.

The state equation for the ideal gas is

$$pv = R\theta$$

or equivalently

$$\frac{v}{R} = \frac{\theta}{p}.$$

Taking derivatives, the above equation becomes

$$p \, \delta v + v \, \delta p = R \, \delta \theta.$$

Under constant specific volume ($\delta v = 0$) we obtain

$$v \, \delta p = R \, \delta \theta$$

and thus

$$\frac{\delta \theta}{\delta p} = \frac{v}{R} = \frac{\theta}{p},$$

where $pv = R\theta$. Using the above equation, the relation between the thermal energy and the specific heat,

$$P \, \delta p = c_v \, \delta \theta,$$

can be rewritten as

$$P = C_v \frac{\delta \theta}{\delta p} = \frac{c_v \theta}{p}.$$

Similarly, take the thermal energy $Q \, \delta v$ (J) that is required to increase the specific volume of δv of a gas of unit mass keeping the pressure constant. Here Q is an unknown constant independent of the pressure variation. Taking the specific heat c_p under constant pressure, we get

$$Q \, \delta v = c_p \, \delta \theta,$$

where $\delta \theta$ denotes the increase of temperature of the gas due to the thermal energy $Q \, \delta v$.

Introducing again the state equation for an ideal gas,

$$pv = R\theta$$

and

$$p \, \delta v + v \, \delta p = R \, \delta \theta,$$

we obtain under constant pressure ($\delta p = 0$)

$$p \, \delta v = R \, \delta \theta$$

or

$$\frac{\delta \theta}{\delta v} = \frac{p}{R} = \frac{\theta}{v},$$

where $pv = R\theta$. Using this equation above, the relation between the thermal energy and the specific heat,

$$Q \, \delta v = c_p \, \delta \theta,$$

can be rewritten as

$$Q = c_p \frac{\delta\theta}{\delta v} = c_p \frac{\theta}{v}.$$

Consequently, we define the thermal energy variation under the condition that both the pressure and specific volume change as

$$P\,\delta p + Q\,\delta v = \theta \left(\frac{c_v\,\delta p}{p} + \frac{c_p\,\delta v}{v} \right).$$

If the process is adiabatic, the thermal energy variation above must be zero. That is,

$$\frac{\delta p}{p} + \frac{c_p\,\delta v}{c_v v} = 0.$$

Integrating both sides of the equation above, we get

$$\ln p + \gamma \ln v = \text{constant}$$

or

$$pv^\gamma = \text{constant},$$

where

$$\gamma = \frac{c_p}{c_v}.$$

The ratio γ is about 1.4 for air.

A gas also has elasticity. The dilatation Δ is defined as

$$\Delta = \frac{\delta v}{v}$$

when the gas has pressure p and volume v at the initial state and these are changed by δp and δv. Similarly, the compression is given as $-\Delta$. The bulk modulus (volume elasticity) is defined as the ratio between the pressure variation and the compression, given by

$$K = \frac{\delta p}{-\Delta} \quad (\text{N m}^{-2})$$

The bulk modulus cannot be determined by the initial and final states, but depends on the variation process.

6.1.2   Plane Wave Equation

We derive the wave equation in air for a 1-D system such as acoustic pipe, following Rossing and Fletcher (1995). Suppose that ξ denotes the displacement (m) of the air during the passage of a sound wave, so that the element ABCD of thickness dx changes to A′ B′ C′ D′ as shown in Fig. 6.1.1. The volume of this element becomes

$$v + dv = S\,dx\left(1 + \frac{\partial\xi}{\partial x}\right) \quad (\mathrm{m}^3)$$

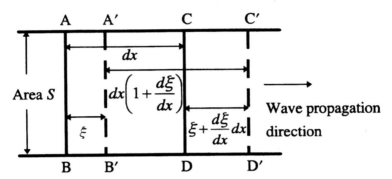

Figure 6.1.1 Displacement of the air during the passage of a sound wave.

or equivalently

$$\frac{dv}{v} = \frac{\partial\xi}{\partial x},$$

where S denotes the area (m^2) normal to x (m),

$$v = S\,dx \quad (\mathrm{m}^3)$$

and

$$dv = S\,dx\frac{\partial\xi}{\partial x} \quad (\mathrm{m}^3).$$

The bulk modulus K (Pa) is defined by the relationship

$$dp = -K\frac{dv}{v} \quad (\mathrm{Pa})$$

where dp denotes the pressure variation. From here we will write dp as p due to acoustic wave pressure (sound pressure). Hence the sound pressure p (Pa) is expressed using the displacement ξ (m):

$$p = -K\frac{dv}{v} = -K\frac{\partial\xi}{\partial x} \quad (\mathrm{Pa}).$$

Next, if we introduce the Newtonian relation of motion between the pressure gradient force in the x direction and the acceleration of motion, we get

$$-S \left(\frac{\partial p}{\partial x} dx \right) = (\rho S \, dx) \frac{\partial^2 \xi}{\partial t^2} \quad \text{(N)}$$

or equivalently

$$-\frac{\partial p}{\partial x} = \rho \frac{\partial^2 \xi}{\partial t^2} \quad (\text{Pa m}^{-1}),$$

where ρ (kg m^{-3}) is the volume density of the air.
  Differentiating

$$p = -K \frac{\partial \xi}{\partial x} \quad \text{(Pa)}$$

which is the relationship between the sound pressure and the displacement, twice with respect to t, we get

$$\frac{\partial^2 p}{\partial t^2} = -K \frac{\partial^2}{\partial t^2} \left(\frac{\partial \xi}{\partial x} \right).$$

And differentiating

$$-\frac{\partial p}{\partial x} = \rho \frac{\partial^2 \xi}{\partial t^2}$$

already derived from the Newtonian equation, with respect to x, we obtain

$$-\frac{\partial^2 p}{\partial x^2} = \rho \frac{\partial}{\partial x} \left(\frac{\partial^2 \xi}{\partial t^2} \right) = \rho \frac{\partial^2}{\partial t^2} \left(\frac{\partial \xi}{\partial x} \right).$$

Consequently, we can get the relationship between the temporal and spatial changes of sound pressure propagation,

$$\frac{\partial^2 p}{\partial t^2} = \frac{K}{\rho} \frac{\partial^2 p}{\partial x^2} = c^2 \frac{\partial^2 p}{\partial x^2} \quad (\text{Pa s}^{-2}),$$

where

$$\frac{\partial}{\partial x} \frac{\partial^2 \xi}{\partial t^2} = -\frac{1}{\rho} \frac{\partial^2 p}{\partial x^2} = -\frac{1}{K} \frac{\partial^2 p}{\partial t^2}$$

and

$$c \equiv \sqrt{\frac{K}{\rho}} \quad (\text{m s}^{-1})$$

denotes the sound speed in the air. This is called the wave equation for plane waves in a 1-D space.

6.1.3 Sound Speed (Lamb, 1945)

As stated in Section 6.1.1, the bulk modulus depends on the thermal process. The bulk modulus and the volume of a gas are related by Boyle's law under isothermal conditions as

$$pv = \text{constant},$$

or

$$\frac{p + dp}{p} = \frac{v}{v + dv}.$$

Neglecting the second-order infinitesimal quantity, $dp\,dv$, we get

$$-\frac{v}{dv}\,dp \cong p$$

since

$$pv + p\,dv + dp\,v + dp\,dv = pv.$$

Thus the bulk modulus is given by

$$K = -\frac{v}{dv}\,dp \cong p \quad \text{(Pa)}.$$

Similarly, if we take the adiabatic condition, then

$$pv^{\gamma} = \text{constant},$$

or

$$\frac{p + dp}{p} = \left(\frac{v}{v + dv}\right)^{\gamma}.$$

Thus we get

$$(p + dp)v^{\gamma}\left(1 + \frac{dv}{v}\right)^{\gamma} = pv^{\gamma},$$

and

$$K = -\frac{v}{dv}\,dp \cong \gamma p,$$

where $\gamma = c_p/c_v$ denotes the ratio of the specific heats of air at constant pressure and at constant volume.

If the sound wave motion propogation in air is isothermal, the sound speed at $0°\text{C}$ becomes

$$c = \sqrt{\frac{K}{\rho}} = \sqrt{\frac{p}{\rho}} = \sqrt{\frac{0.76 \times 13.6 \times 10^3 \times 9.81}{1.29}} = 280 \quad (\text{m}\,\text{s}^{-1})$$

where p is the static barometric pressure (Pa),

$$p = 0.76 \times 13.6 \times 10^3 \times 9.81 \quad \text{(Pa)}$$

and $\rho = 1.29$ $(\text{kg}\,\text{m}^{-3})$. This is called Newton's sound speed. In contrast, if the sound wave propagation is adiabatic, the sound speed at $0°\text{C}$ is given by

$$c = \sqrt{\frac{K}{\rho}} = \sqrt{\frac{\gamma p}{\rho}} = \sqrt{\frac{1.41 \times 0.76 \times 13.6 \times 10^3 \times 9.81}{1.29}}$$

$$= 332 \quad (\text{m s}^{-1})$$

This is called Laplace's sound speed. Sound wave behavior in air for ordinary wavelengths is adiabatic.

6.1.4 General Solutions of the Plane Wave Equation

Possible solutions of the wave equation are composed of two waves traveling in the right and the left directions respectively:

$$p(x, t) = f_1(x - ct) + f_2(x + ct)$$

$$= f_1(u_1) + f_2(u_2) \quad (\text{Pa}),$$

where f_1 and f_2 are arbitrary continuous functions. We can confirm whether this solution satisfies the wave equation

$$\frac{\partial^2 p}{\partial t^2} = c^2 \frac{\partial^2 p}{\partial x^2}.$$

The first and second derivatives of the sound pressure with respect to time become

$$\frac{\partial p}{\partial t} = \frac{\partial p}{\partial u_1} \frac{\partial u_1}{\partial t} + \frac{\partial p}{\partial u_2} \frac{\partial u_2}{\partial t} = \frac{\partial f_1}{\partial u_1}(-c) + \frac{\partial f_2}{\partial u_2}(+c)$$

and

$$\frac{\partial^2 p}{\partial t^2} = \frac{\partial}{\partial u_1} \left(\frac{\partial f_1}{\partial u_1}(-c) \right)(-c) + \frac{\partial}{\partial u_2} \left(\frac{\partial f_2}{\partial u_2}(+c) \right)(+c)$$

$$= c^2 \left(\frac{\partial^2 f_1}{\partial u_1^2} + \frac{\partial^2 f_2}{\partial u_2^2} \right),$$

respectively. Here $u_1 = x - ct$ and $u_2 = x + ct$. Similarly, we obtain the second derivative of the sound pressure with respect to the spatial coordinate x as

$$\frac{\partial^2 p}{\partial x^2} = \frac{\partial^2 f_1}{\partial u_1^2} + \frac{\partial^2 f_2}{\partial u_2^2}.$$

Therefore, we can see that the wave equation holds as

$$\frac{\partial^2 p}{\partial t^2} = c^2 \left(\frac{\partial^2 f_1}{\partial u_1^2} + \frac{\partial^2 f_2}{\partial u_2^2} \right) = c^2 \frac{\partial^2 p}{\partial x^2}.$$

If we assume $\exp(j\omega t)$ time dependency, then the wave equation

$$\frac{\partial^2 p}{\partial t^2} = c^2 \frac{\partial^2 p}{\partial x^2}$$

becomes

$$\frac{\partial^2 p}{\partial x^2} = -\frac{\omega^2}{c^2} p = -k^2 p,$$

where $k = \omega/c = 2\pi/\lambda$ is the wavenumber and λ is the wavelength (m). A general solution for the $\exp(j\omega t)$ time-dependent wave equation

$$\frac{\partial^2 p}{\partial x^2} = -k^2 p$$

is given by

$$p = Ae^{j\omega t - jkx} + Be^{j\omega t + jkx} \quad \text{(Pa)},$$

where the A and B terms travel to the right and left, respectively, and kx denotes the phase change of the waves between two positions with separation x. The phase of the progressive wave changes by kx while the wave travels over a distance x.

6.1.5 Specific Impedance

As stated in Section 6.1.1, the relationship between the gradient of the sound pressure and the particle acceleration is defined by

$$-\frac{\partial p}{\partial x} = \rho \frac{\partial^2 \xi}{\partial t^2}.$$

If we substitute progressive sinusoidal pressure waves by

$$p = p_0 e^{j\omega t - jkx} \quad \text{and} \quad \xi = \xi_0 e^{j\omega t - jkx},$$

we obtain

$$-\frac{\partial p}{\partial x} = jkp = \rho \frac{\partial^2 \xi}{\partial t^2} = \rho j\omega \frac{\partial \xi}{\partial t} = j\omega \rho v,$$

where we set $v = \partial \xi / \partial t$. Thus the particle velocity v for the sinusoidal progressive wave is derived from the pressure as

$$v = \frac{k}{\rho \omega} p = \frac{1}{\rho c} p \quad \text{(m s}^{-1})$$

or equivalently

$$p = \rho c v \quad (\text{Pa}).$$

The acoustic impedance is defined as the ratio of the sound pressure (Pa) to the particle velocity (m s^{-1}) as

$$z \equiv \frac{p}{v} \quad (\text{Pa s m}^{-1}).$$

In particular, for a sinusoidal progressive wave given by $p = p_0 e^{j\omega t - jkx}$, the acoustic impedance becomes

$$z_0 = \frac{p}{v} = \rho c \quad (\text{Pa s m}^{-1})$$

This impedance z_0 is called the specific impedance and is a property of the medium. The specific impedance of a medium is the ratio between the sound pressure and the particle velocity of a progressive plane wave traveling in the medium. If we take 340 (m s^{-1}) as the sound speed in air with a volume density of 1.3 (kg m^{-3}), the specific impedance becomes 442 (Pa s m^{-1}).

6.1.6 General Wave Equation (Lamb, 1945)

Sound waves do not always propagate as plane waves. In this section we derive the general wave equation following Lamb in *Hydrodynamics*. Suppose that we take a small rectangular element having its center at (x, y, z) and its edge δx, δy, δz parallel to the rectangular coordinate axis. The volume of the element is $\delta x \, \delta y \, \delta z$ and the mass is $\rho_0 \, \delta x \, \delta y \, \delta z$. We form the Newtonian equation of motion by regarding this small element as a particle which has a mass of $\rho_0 \, \delta x \, \delta y \, \delta z$.

Let u, v, w be the components, parallel to the coordinate axes, of the velocity at the point (x, y, z) at time t. These quantities are then functions of the independent variables x, y, z, t. The acceleration of the x-component of the momentum of the particle is given by

$$\frac{\partial}{\partial t}[(\rho_0 \, \delta x \, \delta y \, \delta z)u] = (\rho_0 \, \delta x \, \delta y \, \delta z)\frac{\partial u}{\partial t},$$

where the time derivative of the density ρ_0 is assumed to be zero since the wave motion is sufficiently slow. This acceleration must be equal to the x-component of the forces acting on the element.

The fluid element is so small that the mean pressure over any face of this small element may be taken to be equal to the pressure at the center of that surface. The pressure on the yz^--face which is located at $x - \frac{1}{2}\delta x$ becomes

$$p_{yz^-} = p - \frac{1}{2}\frac{\partial p}{\partial x}\delta x,$$

where p denotes the pressure at the center of the element. The pressure on the opposite surface becomes

$$p_{yz^+} = p + \frac{1}{2}\frac{\partial p}{\partial x}\delta x.$$

The difference of these gives

$$p_{yz^-} - p_{yz^+} = -\frac{\partial p}{\partial x}\delta x$$

in the direction of x-positive. Then the x-component of the forces acting on the element is given by

$$F_x = -\frac{\partial p}{\partial x}\delta x\,\delta y\,\delta z.$$

Similarly, we can obtain the forces for other components as

$$F_y = -\frac{\partial p}{\partial y}\delta x\,\delta y\,\delta z \quad \text{for the } y\text{-component}$$

and

$$F_z = -\frac{\partial p}{\partial z}\delta x\,\delta y\,\delta z \quad \text{for the } z\text{-component.}$$

Consequently, we have the Newtonian equation of motion as

$$\rho_0\frac{\partial u}{\partial t} = -\frac{\partial p}{\partial x},$$

$$\rho_0\frac{\partial v}{\partial t} = -\frac{\partial p}{\partial y},$$

$$\rho_0\frac{\partial w}{\partial t} = -\frac{\partial p}{\partial z},$$

or equivalently

$$\rho_0\frac{\partial \mathbf{V}}{\partial t} = -\nabla p,$$

where $\mathbf{V}$ denotes the velocity vector.

Next we consider the pressure changes due to the sound wave traveling. As already described in Section 6.1.2, the sound pressure produced by the passage of the sound wave can be related to the bulk modulus of the medium in which the sound wave travels. The sound pressure is sufficiently small so that it can be taken to be proportional to the changes in density. The sound pressure can be written as

$$p = -K \frac{\delta v}{v_0} \cong K \frac{\delta \rho}{\rho_0} \quad \text{(Pa)},$$

where K denotes the bulk modulus or volume elasticity, and v_0 and ρ_0 are respectively the volume and the volume density in the undisturbed state.

The approximation above can be confirmed by neglecting the second-order infinitesimal quantity as follows. The change of the density $\delta\rho$ is written as

$$\rho = \rho_0 + \delta\rho = \rho_0 \left(1 + \frac{\delta\rho}{\rho_0}\right) = \rho_0(1 + s),$$

where $s = \delta\rho/\rho_0$ is called the condensation. Similarly, the change in the volume is described by

$$v = v_0 + \delta v = v_0 \left(1 + \frac{\delta v}{v_0}\right) = v_0(1 + \Delta),$$

where $\Delta = \delta v/v_0$ is called the dilatation. Since the total mass is not changed, we have the relationship

$$\rho v = \rho_0(1 + s)v_0(1 + \Delta) = \rho_0 v_0,$$

or equivalently

$$(1 + s)(1 + \Delta) = 1.$$

Neglecting the second-order quantity, we get

$$s \cong -\Delta.$$

Introducing this condensation, the sound pressure can be written as

$$p = -K \frac{\delta v}{v_0} \cong K \frac{\delta \rho}{\rho_0} = Ks.$$

Substituting this relation into the equation of motion

$$\rho_0 \frac{\partial \mathbf{V}}{\partial t} = -\nabla p$$

we can get the equation

$$\rho_0 \frac{\partial \mathbf{V}}{\partial t} = -K \nabla s,$$

where we assume $\rho \cong \rho_0$. Integrating with respect to t, we have

$$\mathbf{V} = -c^2 \nabla \left(\int_0^t s \, dt + \mathbf{V}_0 \right) \equiv -\nabla \phi,$$

where $\mathbf{V}_0$ is the velocity at the point (x,y,z) at the instant $t = 0$ and $c^2 = K/\rho_0$.

In the equation above we have introduced a scalar quantity ϕ which is defined by

$$\phi = c^2 \int_0^t s \, dt + \phi_0,$$

or equivalently

$$\frac{\partial \phi}{\partial t} = c^2 s.$$

This is called the velocity potential, whose existence is proved in an irrotational motion field (Lamb, 1945). We can write the relationship between the sound pressure and the velocity potential as well as that for the particle velocity. Recalling the relation

$$p = -K \frac{\delta v}{v_0} \cong K \frac{\delta \rho}{\rho_0} = Ks$$

we can rewrite the sound pressure using the velocity potential as

$$p = K \frac{1}{c^2} \frac{\partial \phi}{\partial t} = \rho_0 \frac{\partial \phi}{\partial t}.$$

We require one more equation, which is called the equation of continuity in fluid dynamics, in order to get the general wave equation. We have to calculate the small amount of change of the mass included in the element during the unit time interval of fluid motion. Again take a small rectangular element having its center at (x, y, z) and its edges δx, δy, δz parallel to the rectangular coordinate axis. The amount of matter which enters the element across the yz^--face per unit time is given by

$$\delta \rho_{yz^-} = \left(\rho u - \frac{1}{2} \frac{\partial(\rho u)}{\partial x} \delta x \right) \delta y \, \delta z$$

and the amount which leaves the element by the opposite face is

$$\delta \rho_{yz^+} = \left(\rho u + \frac{1}{2} \frac{\partial(\rho u)}{\partial x} \delta x \right) \delta y \, \delta z.$$

The two faces together give a gain of

$$\delta\rho_{yz-} - \delta\rho_{yz+} = -\frac{\partial(\rho u)}{\partial x}\,\delta x\,\delta y\,\delta z$$

per unit time.

Similarly, we can calculate the effect of the flux across the remaining faces. Consequently, we have the total gain per unit time in the space $\delta x\,\delta y\,\delta z$ as

$$\delta\rho = -\left[\frac{\partial(\rho u)}{\partial x} + \frac{\partial(\rho v)}{\partial y} + \frac{\partial(\rho w)}{\partial z}\right]\delta x\,\delta y\,\delta z.$$

This gain must be equal to the change of mass in the element as

$$\frac{\partial}{\partial t}(\rho\,\delta x\,\delta y\,\delta z) = -\left[\frac{\partial(\rho u)}{\partial x} + \frac{\partial(\rho v)}{\partial y} + \frac{\partial(\rho w)}{\partial z}\right]\delta x\,\delta y\,\delta z,$$

or equivalently

$$\frac{\partial\rho}{\partial t} + \left[\frac{\partial(\rho u)}{\partial x} + \frac{\partial(\rho v)}{\partial y} + \frac{\partial(\rho w)}{\partial z}\right] = 0,$$

if there are no sinks and sources in the element. This is called the *equation of continuity*.

Now we can obtain the general wave equation. We already have two equations:

$$\rho_0\frac{\partial \mathbf{V}}{\partial t} = -\nabla p \qquad \text{equation of motion}$$

and

$$\frac{\partial\rho}{\partial t} + \left[\frac{\partial(\rho u)}{\partial x} + \frac{\partial(\rho v)}{\partial y} + \frac{\partial(\rho w)}{\partial z}\right] = 0 \qquad \text{equation of continuity.}$$

We have also derived the next equation from the equation of motion by introducing the velocity potential ϕ,

$$\frac{\partial\phi}{\partial t} = c^2 s$$

where

$$\mathbf{V} = -\nabla\phi, \qquad p = \rho_0\frac{\partial\phi}{\partial t},$$

and s denotes the condensation and $\mathbf{V}$ is the particle velocity vector.

By writing $\rho = \rho_0(1 + s)$, we can rewrite the equation of continuity as

$$\rho_0 \frac{\partial s}{\partial t} + \left[\rho \frac{\partial u}{\partial x} + u \frac{\partial \rho}{\partial x} + \rho \frac{\partial v}{\partial y} + v \frac{\partial \rho}{\partial y} + \rho \frac{\partial w}{\partial z} + w \frac{\partial \rho}{\partial z} \right]$$

$$= \rho_0 \frac{\partial s}{\partial t} + \left[-\rho \frac{\partial^2 \phi}{\partial x^2} + u \frac{\partial \rho}{\partial x} - \rho \frac{\partial^2 \phi}{\partial y^2} + v \frac{\partial \rho}{\partial y} - \rho \frac{\partial^2 \phi}{\partial z^2} + w \frac{\partial \rho}{\partial z} \right]$$

$$\cong \rho_0 \frac{\partial s}{\partial t} + \left[-\rho_0 \frac{\partial^2 \phi}{\partial x^2} - \rho_0 \frac{\partial^2 \phi}{\partial y^2} - \rho_0 \frac{\partial^2 \phi}{\partial z^2} \right]$$

$$\cong 0,$$

or

$$\frac{\partial s}{\partial t} \cong \frac{\partial^2 \phi}{\partial x^2} + \frac{\partial^2 \phi}{\partial y^2} + \frac{\partial^2 \phi}{\partial z^2}.$$

Introducing

$$\frac{1}{c^2} \frac{\partial \phi}{\partial t} = s$$

we get the wave equation

$$\frac{\partial^2 \phi}{\partial t^2} = c^2 \left(\frac{\partial^2 \phi}{\partial x^2} + \frac{\partial^2 \phi}{\partial y^2} + \frac{\partial^2 \phi}{\partial z^2} \right) = c^2 \nabla^2 \phi.$$

If we change the variable from the velocity potential to the sound pressure using

$$p = \rho_0 \frac{\partial \phi}{\partial t},$$

we get the wave equation for the sound pressure:

$$\frac{\partial^2 p}{\partial t^2} = c^2 \nabla^2 p.$$

6.1.7 Helmholtz Equation and Spherical Wave

The wave equation for 1-D space can be extended into 3-D space by

$$\frac{\partial^2 p}{\partial t^2} = c^2 \nabla^2 p.$$

Assuming $\exp(j\omega t)$ time dependency, we get

$$\nabla^2 p = -k^2 p,$$

where

$$\nabla^2 p = \frac{\partial^2 p}{\partial x^2} + \frac{\partial^2 p}{\partial y^2} + \frac{\partial^2 p}{\partial z^2}$$

in rectangular coordinate notation. This equation is called the Helmholtz equation.

The simplest wave propagation in 3-D space is a spherical wave which satisfies the wave equation in spherical coordinates,

$$\nabla^2 p = \frac{1}{r^2}\frac{\partial}{\partial r}\left(r^2\frac{\partial p}{\partial r}\right) = -k^2 p.$$

Taking a new variable as $\psi = rp$, we can rewrite the equation above as

$$\nabla^2 p = \frac{1}{r^2}\frac{\partial}{\partial r}\left(r^2\frac{\partial p}{\partial r}\right) = \frac{1}{r}\frac{\partial^2\psi}{\partial r^2} = -k^2\frac{\psi}{r},$$

thus we get

$$\frac{\partial^2\psi}{\partial r^2} + k^2\psi = 0,$$

where

$$p = \frac{\psi}{r}, \qquad r^2\frac{\partial p}{\partial r} = r^2\left(\frac{\partial\psi}{\partial r}\frac{1}{r} - \frac{\psi}{r^2}\right) = r\frac{\partial\psi}{\partial r} - \psi,$$

and

$$\frac{1}{r^2}\frac{\partial}{\partial r}\left(r^2\frac{\partial p}{\partial r}\right) = \frac{1}{r^2}\frac{\partial}{\partial r}\left(r\frac{\partial\psi}{\partial r} - \psi\right) = \frac{1}{r^2}\left(\frac{\partial\psi}{\partial r} + r\frac{\partial^2\psi}{\partial r^2} - \frac{\partial\psi}{\partial r}\right)$$

$$= \frac{1}{r}\frac{\partial^2\psi}{\partial r^2}.$$

The general solution for the wave equation

$$\frac{\partial^2\psi}{\partial r^2} + k^2\psi = 0$$

is written as

$$p = \left(\frac{A}{r}e^{-jkr} + \frac{B}{r}e^{jkr}\right)e^{j\omega t} \quad (\text{Pa}),$$

which is a superposition of an outgoing ($B = 0$) and an incoming ($A = 0$) spherical wave and $\psi = rp$.

The equation for the 1-D case,

$$-\frac{\partial p}{\partial x} = \rho\frac{\partial v}{\partial t}$$

can be rewritten for the 3-D case as

$$-\nabla p = -\frac{\partial p}{\partial r} = \rho\frac{\partial v}{\partial t},$$

where v denotes the particle velocity ($m\,s^{-1}$), and the sound pressure p depends only on r and t. The acoustic impedance, which is defined by the ratio of the sound pressure and the particle velocity for an outgoing spherical wave with angular frequency ω, is

$$z = \frac{p}{v} = \rho c \left(\frac{jkr}{1 + jkr} \right) \quad (Pa\,s\,m^{-1}).$$

This can be formulated as

$$-\nabla p = -\frac{\partial p}{\partial r} = \rho \frac{\partial v}{\partial t} = j\omega\rho v = j\rho ckv$$

and

$$-\frac{\partial p}{\partial r} = -\frac{\partial}{\partial r} \left(\frac{A}{r} e^{-jkr} \right) = \frac{1 + jkr}{r} p.$$

The pressure is not in phase with the particle velocity; in a plane wave there are no phase differences between the pressure and velocity.

A simple source which radiates a spherical wave is a point source. The velocity potential of the radiated sound field from a point source can be written in the form

$$\phi = \frac{A}{r} e^{j\omega t - jkr},$$

where r denotes the distance from the source and ω is the angular frequency of the source. Here the coefficient A is determined by the strength of the source ($m^3\,s^{-1}$).

Suppose that the volume velocity of the source is $Qe^{j\omega t}$ ($m^3\,s^{-1}$). We can describe the relationship on the surface of the source as

$$4\pi a^2 \frac{-\partial \phi}{\partial r} \bigg|_{r=a} = 4\pi a^2 A \frac{1 + jkr}{r^2} e^{j\omega t - jkr} \bigg|_{r=a}$$

$$= 4\pi A (1 + jka) e^{j\omega t - jka} \equiv Qe^{j\omega t}.$$

Thus, the coefficient A is given by

$$A = \frac{Qe^{jka}}{4\pi(1 + jka)} \quad (m^3\,s^{-1}).$$

If we take the limit as the radius a approaches zero, we can get the coefficient A for a point source,

$$A = \lim_{a \to 0} \frac{Qe^{jka}}{4\pi(1 + jka)} = \frac{Q}{4\pi}.$$

The radiation field from the point source becomes

$$\phi = \frac{A}{r}e^{j\omega t - jkr} = \frac{Q}{4\pi r}e^{j\omega t - jkr},$$

where the point source has the volume velocity or strength of Q $(m^3 \, s^{-1})$.

Another simple example of sources is a double source. Suppose that we have a pair of point sources with opposite phase which are closely located as shown in Fig. 6.1.2. In Fig. 6.1.2, we assume the relation

$$m|\mathbf{R}| = \lim_{R \to 0} mR \equiv \mu = \text{const.},$$

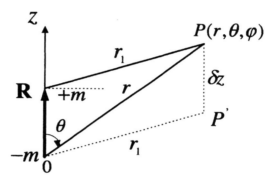

Figure 6.1.2 A double source and its axis.

where m is the strength of each point source, and the vector $\mathbf{R}$ and μ are called the axis and strength of the double source, respectively.

The velocity potential at the point P in Fig. 6.1.2 can be written as

$$\phi_P = \frac{m}{4\pi}\left(\frac{e^{-jkr_1}}{r_1} - \frac{e^{-jkr}}{r}\right).$$

Here, the velocity potential at the point P' in Fig. 6.1.2 due to the point source of $-m$, is given by

$$\phi_{(P' \leftarrow -m)} = \frac{-me^{-jkr_1}}{4\pi r_1}.$$

Thus, we can rewrite

$$\phi_P = \frac{m}{4\pi}\left(\frac{e^{-jkr_1}}{r_1} - \frac{e^{-jkr}}{r}\right) = \frac{-m}{4\pi}\left(\frac{-e^{-jkr_1}}{r_1} + \frac{e^{-jkr}}{r}\right)$$

$$= \frac{-m}{4\pi} \left(\frac{e^{-jkr}}{r} - \frac{e^{-jkr_1}}{r_1} \right) = \phi_{(P \leftarrow -m)} - \phi_{(P' \leftarrow -m)}$$

$$\rightarrow \frac{-mR}{4\pi} \frac{\partial}{\partial z} \left(\frac{e^{-jkr}}{r} \right) = \frac{-\mu}{4\pi} \frac{\partial}{\partial z} \left(\frac{e^{-jkr}}{r} \right) \qquad (R \rightarrow 0).$$

Changing variables as shown in Fig. 6.1.2 such that

$$\delta r = \delta z \cos\theta \qquad \text{and} \qquad \frac{\partial}{\partial z} = \frac{\partial}{\partial r} \cos\theta,$$

we can rewrite the equation above for the velocity potential radiated from a double source as

$$\phi_P = \frac{-\mu}{4\pi} \frac{\partial}{\partial z} \left(\frac{e^{-jkr}}{r} \right) = \frac{-\mu}{4\pi} \frac{\partial}{\partial r} \left(\frac{e^{-jkr}}{r} \right) \cos\theta$$

$$= \frac{\mu \cos\theta}{4\pi r} \left(\frac{1 + jkr}{r} \right) e^{-jkr} \rightarrow \begin{cases} \dfrac{\mu \cos\theta}{4\pi r^2} & (kr \rightarrow 0) \\[2mm] \dfrac{jk\mu \cos\theta}{4\pi r} e^{-jkr} & (kr \gg 1). \end{cases}$$

6.1.8 Reflection of a Plane Wave at a Boundary

One of the fundamental wave propagation problems is reflection at a boundary. Let us take a coordinate as shown in Fig. 6.1.3 and define the distance of the sound wave path as ζ, whose positive direction is taken from the origin O which is the wave incidence position. A plane wave travels in the xz-plane in medium 1 and is incident to the boundary (xy-plane ($z = 0$)) between medium 1 and medium 2 at an angle of incidence of θ_i. Suppose that the reflected wave goes back at an angle of θ_r into medium 1 and the transmitted wave travels through the medium 2 at an angle of θ_t.

We can write these three waves using the velocity potential as

$$\phi_i = \phi_{i0} e^{j\omega t + jk_1 \zeta_i} \qquad \text{for the incident wave,}$$

$$\phi_r = \phi_{r0} e^{j\omega t - jk_1 \zeta_r} \qquad \text{for the reflected wave,}$$

and

$$\phi_t = \phi_{t0} e^{j\omega t - jk_2 \zeta_t} \qquad \text{for the transmitted wave,}$$

respectively, where

$$\zeta_i = -x \sin\theta_i + z \cos\theta_i,$$

$$\zeta_r = x \sin\theta_r + z \cos\theta_r,$$

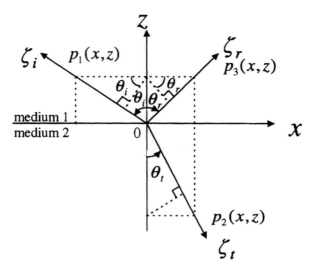

Figure 6.1.3   Incident, reflected, and transmitted waves at the boundary of two media.

and

$$\zeta_t = x \sin \theta_t - z \cos \theta_t,$$

Using the expressions above, we can write the particle velocity and the sound pressure as follows. The z-component of the particle velocity becomes

$$v_{iz} = -\frac{\partial \phi_i}{\partial z} = -j\phi_i k_1 \cos \theta_i \qquad \text{for the incident wave,}$$

$$v_{rz} = -\frac{\partial \phi_r}{\partial z} = j\phi_r k_1 \cos \theta_r \qquad \text{for the reflected wave,}$$

and

$$v_{tz} = -\frac{\partial \phi_t}{\partial z} = -j\phi_t k_2 \cos \theta_t \qquad \text{for the transmitted wave.}$$

Similarly, we have the sound pressure as

$$p_i = \rho_1 \frac{\partial \phi_i}{\partial t} = j\omega\rho_1\phi_i \qquad \text{for the incident wave,}$$

$$p_r = \rho_1 \frac{\partial \phi_r}{\partial t} = j\omega\rho_1\phi_r \qquad \text{for the reflected wave,}$$

and

$$p_t = \rho_2 \frac{\partial \phi_t}{\partial t} = j\omega\rho_2\phi_t \qquad \text{for the transmitted wave.}$$

We impose boundary conditions on the boundary ($z = 0$) as

$$\phi_{i0}e^{-jk_1x\sin\theta_i}k_1\cos\theta_i - \phi_{r0}e^{-jk_1x\sin\theta_r}k_1\cos\theta_r$$
$$= \phi_{t0}e^{-jk_2x\sin\theta_t}k_2\cos\theta_t,$$

(continuity of z-component of the particle velocity at $z = 0$) and

$$\rho_1\phi_{i0}e^{-jk_1x\sin\theta_i} + \rho_1\phi_{r0}e^{-jk_1x\sin\theta_r} = \rho_2\phi_{t0}e^{-jk_2x\sin\theta_t}$$

(continuity of the sound pressure at $z = 0$). These boundary conditions must hold well for any x. Thus, we require the next relation:

$$e^{-jk_1x\sin\theta_i} = e^{-jk_1x\sin\theta_r} = e^{-jk_2x\sin\theta_t}$$

or

$$k_1\sin\theta_i = k_1\sin\theta_r = k_2\sin\theta_t.$$

That is, the reflection angle must be equal to the incidence angle.

Consequently, we can rewrite the boundary conditions as

$$\phi_{i0} - \phi_{r0} = \frac{c_1\cos\theta_t}{c_2\cos\theta_i}\phi_{t0}$$

and

$$\phi_{i0} + \phi_{r0} = \frac{\rho_2}{\rho_1}\phi_{t0}$$

where $k_1 = w/c_1$ and $k_2 = w/c_2$

Then we can get the reflection coefficient of the sound pressure as

$$R \equiv \frac{p_r}{p_i}\bigg|_{z=0} = \frac{\phi_r}{\phi_i}\bigg|_{z=0} = \frac{\phi_{r0}}{\phi_{i0}} = \frac{z_2\cos\theta_i - z_1\cos\theta_t}{z_1\cos\theta_t + z_2\cos\theta_i} = \frac{\beta\cos\theta_i - \cos\theta_t}{\beta\cos\theta_i + \cos\theta_t},$$

where $z_1 = \rho_1c_1$, $z_2 = \rho_2c_2$, and $\beta = z_2/z_1$. When β becomes infinity (rigid boundary), R becomes unity, and conversely, when β is zero (so-called free boundary), R becomes -1.

6.1.9 Image Theory

We have considered reflection at a boundary for a plane wave. In this section we describe reflection for a spherical wave. The sound field radiated from the point source near the boundary can be analyzed using image theory. Suppose that we have a point source of angular frequency ω located at S near the boundary as shown in Fig. 6.1.4. First we consider a case when the boundary is rigid.

Take a point O as an observation position. If there is no boundary, the velocity potential at O becomes

$$\phi_1 = \frac{A}{4\pi r_1}e^{-jkr_1},$$

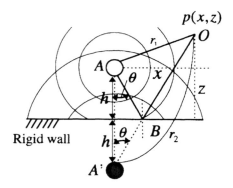

Figure 6.1.4   Sound source, rigid wall, and virtual source.

where A denotes the strength of the source $(m^3 s^{-1})$ and the distance between the receiving and source positions is r_1. If we have a boundary as shown in Fig. 6.1.4, we have to take account of the reflection wave from the boundary. The reflection wave can be expressed by a spherical wave radiated from a virtual source as shown in Fig. 6.1.4. This virtual source is called an *image source*.

The resultant sound field becomes

$$\phi = \frac{A}{4\pi r_1}e^{-jkr_1} + \frac{A}{4\pi r_2}e^{-jkr_2},$$

where r_2 is the distance between the observation point and the virtual source. When the boundary has reflection coefficient R, the resultant sound field becomes

$$\phi = \frac{A}{4\pi r_1}e^{-jkr_1} + \frac{RA}{4\pi r_2}e^{-jkr_2}.$$

Thus when the boundary is free $(R = -1)$, the virtual source becomes antiphase to the real source. We assume that R is a real quantity when using image theory.

6.1.10   Sound Power Output of a Point Source

The sound power is the sound energy radiated by a source per unit time. Suppose that we have a point source of the strength Q $(m^3 s^{-1})$ in a field. The sound power radiated from the point source can be expressed as

$$P \equiv \tfrac{1}{2}\,\mathrm{Re}\int_\Sigma pv^*\,dS \quad \text{(W)}$$

where Σ is a spherical surface of radius r centered at the point source. The sound pressure (Pa) and radial particle velocity (m s^{-1}) are

$$p = \rho \frac{\partial \phi}{\partial t} = \frac{j\omega\rho Q}{4\pi r} e^{j\omega t - jkr} \quad (\text{Pa})$$

and

$$v = -\frac{\partial \phi}{\partial r} = \frac{Q}{4\pi} \frac{1 + jkr}{r^2} e^{j\omega t - jkr}, \quad (\text{m s}^{-1})$$

respectively, where $*$ denotes the complex conjugate and Re stands for the real part. The sound power output of a point source is given by

$$
\begin{aligned}
P &= \frac{1}{2} \text{Re} \int_\Sigma p v^* \, dS \\
&= \frac{1}{2} \text{Re} \int_\Sigma \left(\frac{j\omega\rho Q}{4\pi r} e^{j\omega t - jkr} \right) \left(\frac{Q}{4\pi} \frac{1 - jkr}{r^2} e^{-j\omega t + jkr} \right) dS \\
&= \frac{1}{2} \left(\frac{Q}{4\pi} \right)^2 \text{Re} \int_\Sigma \left(\frac{j\omega\rho Q}{r} e^{j\omega t - jkr} \right) \left(\frac{1 - jkr}{r^2} e^{-j\omega t + jkr} \right) dS \\
&= \frac{\rho c}{2} \left(\frac{Qk}{4\pi} \right)^2 \int_\Sigma \left(\frac{1}{r^2} \right) dS = \frac{\rho c}{2} \left(\frac{Qk}{4\pi} \right)^2 \left(\frac{4\pi r^2}{r^2} \right) \\
&= \frac{\rho c Q^2 k^2}{8\pi} \quad (\text{W}).
\end{aligned}
$$

When we express the sound pressure and particle velocity using complex quantities, the power quantity is also given by a complex function of time. Similarly to electrical circuit theory, real part of the complex power quantity shows power consumed by a system of our interest. When a point source radiates sound into the surrounding environment, such as air, the vibrational energy of the source is consumed by conversion of the source mechanical vibration energy into air vibration.

We take the complex conjugate and multiply by one-half in calculating the power in order to give the time average. Suppose that electric voltage and current are given by real quantities

$$\varepsilon = E \cos(\omega t - \alpha) \quad (V) \quad \text{and} \quad i = I \cos(\omega t - \beta) \quad (A),$$

respectively. The power consumed in the circuit is expressed as

$$
\begin{aligned}
P &= \frac{1}{T} \int_0^T \varepsilon i \, dt = \frac{1}{T} \int_0^T E \cos(\omega t - \alpha) I \cos(\omega t - \beta) \, dt \\
&= EI \frac{1}{T} \int_0^T \cos(\omega t - \alpha) \cos(\omega t - \beta) \, dt
\end{aligned}
$$

$$= EI \frac{1}{2T} \int_0^T [\cos(2\omega t - \alpha - \beta) + \cos(-\alpha + \beta)] \, dt$$

$$= \frac{EI}{2} \cos(-\alpha + \beta) \quad \text{(W)},$$

where $T = 2\pi/\omega$ denotes the period of the voltage and current, and $\cos(-\alpha + \beta)$ is called the *power efficiency*. If the phase difference between the voltage and current is $\pi/2$, no power is consumed, such as in an *LC* circuit.

Now let us take complex forms as

$$\varepsilon = E e^{j\omega t - j\alpha} \quad \text{and} \quad i = I e^{j\omega t - j\beta},$$

respectively. If we define the power as

$$P = \mathrm{Re} \, \frac{1}{T} \int_0^T \varepsilon i \, dt,$$

then we have

$$P = \mathrm{Re} \, \frac{1}{T} \int_0^T \varepsilon i \, dt = \mathrm{Re} \, \frac{1}{T} \int_0^T E e^{j\omega t - j\alpha} I e^{j\omega t - j\beta} \, dt$$

$$= \mathrm{Re} \, \frac{EI}{T} \int_0^T e^{j2\omega t - j\alpha - j\beta} \, dt = \frac{EI}{T} \int_0^T \cos(2\omega t - \alpha - \beta) \, dt = 0,$$

where $T = 2\pi/\omega$ denotes the period of the voltage. We take the complex conjugate as

$$P = \mathrm{Re} \, \frac{1}{T} \int_0^T \varepsilon i^* \, dt = \mathrm{Re} \, \frac{1}{T} \int_0^T E e^{j\omega t - j\alpha} I e^{-j\omega t + j\beta} \, dt$$

$$= \mathrm{Re} \, \frac{EI}{T} \int_0^T e^{-j(\alpha - \beta)} \, dt = EI \cos(\alpha - \beta).$$

This results requires multiplication by $\frac{1}{2}$ in order to get the real power. Therefore, we finally define the real power using complex quantities as

$$P = \frac{1}{2} \mathrm{Re} \, \frac{1}{T} \int_0^T \varepsilon i^* \, dt.$$

6.1.11   Driving Point Impedance and Sound Power Output

Now let us return to the sound power radiated from a source, which is given by

$$P = \tfrac{1}{2} \mathrm{Re} \int_\Sigma p v^* \, dS \quad \text{(W)}.$$

We will introduce driving point acoustic (or mechanical) impedance, which represents characteristics of the sound power output of a source in order to evaluate the effects of reflecting planes, as we describe in the next section.

We have two kinds of driving point impedance. We can use whichever one is suitable to each problem. The acoustic and mechanical driving point impedances are defined as

$$Z_A = \frac{\text{sound pressure at the source}}{\text{volume velocity of the source}}$$

for acoustic impedance (Pa s m^{-3}), and

$$Z_M = \frac{\text{reacting force at the source by radiated sound}}{\text{vibrating velocity of the source}}$$

for mechanical impedance (N s m^{-1}), respectively.

The sound pressure radiated from a small spherical source of the strength $Qe^{j\omega t} (\text{m}^3 \text{s}^{-1})$ whose radius is a (m) becomes

$$p(r = a) = \rho \frac{\partial \phi}{\partial t}\bigg|_{r=a} = j\omega\rho A \frac{e^{j\omega t - jkr}}{r}\bigg|_{r=a}$$

$$= j\omega\rho \frac{Qe^{jka}}{4\pi(1 + jka)} \frac{e^{j\omega t - jkr}}{r}\bigg|_{r=a}$$

$$= j\omega\rho \frac{Qe^{jka}}{4\pi a(1 + jka)} e^{j\omega t - jka}$$

$$= j\omega\rho \frac{Qe^{j\omega t}}{4\pi a(1 + jka)} \quad (\text{Pa})$$

at the surface where the velocity potential is given by $\phi = (A/r)e^{j\omega t - jkr}$, and $A = (Qe^{jka}/4\pi(1 + jka))$. The acoustic driving point impedance for the source is given by

$$z_A = j\omega\rho \frac{Qe^{j\omega t}}{4\pi a(1 + jka)} \bigg/ Qe^{j\omega t} = \frac{j\omega\rho}{4\pi a(1 + jka)}$$

$$= \frac{j\omega\rho(1 - jka)}{4\pi a(1 + k^2 a^2)} = \frac{j\rho ck(1 - jka)}{4\pi a(1 + k^2 a^2)}$$

$$= \frac{\rho c k^2 a^2}{4\pi a^2(1 + k^2 a^2)} + \frac{j\rho cka}{4\pi a^2(1 + k^2 a^2)} = R_A + jI_A \quad (\text{Pa s m}^{-3}).$$

The sound power output from a source can be estimated by setting the closed surface at which the integral is taken just on the surface of the source. That is,

$$P = \frac{1}{2}\,\mathrm{Re}\int_{\Sigma} pv^* \, dS = \frac{1}{2}\,\mathrm{Re}\int_{\text{Source surface}} pv^* \, dS$$

$$= \frac{1}{2}\,\mathrm{Re}\int_{\text{Source surface}} Qe^{j\omega t} z_A v^* \, dS$$

$$= \frac{1}{2}\,\mathrm{Re}\int_{\text{Source surface}} Qe^{j\omega t}(R_A + jI_A)v^* \, dS$$

$$= \frac{1}{2}\,\mathrm{Re}\left[Qe^{j\omega t}(R_A + jI_A)\int_{\text{Source surface}} v^* \, dS\right]$$

$$= \frac{1}{2}\,\mathrm{Re}[Qe^{j\omega t}(R_A + jI_A)Qe^{-j\omega t}]$$

$$= \frac{1}{2}\,\mathrm{Re}[Q^2(R_A + jI_A)]$$

$$= \frac{Q^2}{2}R_A\,(W)$$

where

$$\int_{\text{Source surface}} v \, dS = Qe^{j\omega t}.$$

If we take the limit as the radius of a spherical source, we get the power output of a point source as

$$P_0 \equiv \lim_{a \to 0} \frac{Q^2}{2} R_A = \lim_{a \to 0} \frac{Q^2}{2}\frac{\rho c k^2 a^2}{4\pi a^2(1 + k^2 a^2)} = \lim_{a \to 0} \frac{\rho c Q^2}{8\pi a^2}\frac{k^2 a^2}{1 + k^2 a^2}$$

$$= \frac{\rho c Q^2 k^2}{8\pi}\quad (W)$$

and

$$R_A \to \frac{\rho c k^2}{4\pi} \equiv R_0 \qquad (a \to 0)$$

although the imaginary part of the driving point impedance has no limit. The sound power output can be estimated by the real part of the driving point impedance.

6.1.12   Sound Power Output of a Point Source over a Reflecting Plane (Waterhouse, 1963)

The sound power of a source is a primary concern for sound space control. The sound power output from a source, however, varies owing to environmental conditions, and differs from the free-field value. Estimation of the sound power of a point source over a reflecting plane can be done using image theory.

Suppose that a point source whose volume velocity is $Qe^{i\omega t}$ (m^3 s^{-1}) is located over a reflecting plane as shown in Fig. 6.1.5. If we set an image source at point S' shown in Fig. 6.1.5, we can express the sound pressure at the original point source position of S as

$$p(S) = Qe^{j\omega t}Z_{A0} + j\omega\rho\frac{Q}{4\pi 2h}e^{j\omega t - 2jkh} \quad \text{(Pa)},$$

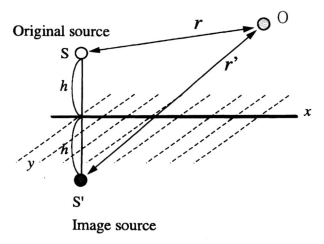

Original source

Figure 6.1.5 A point source over a reflecting plane.

where $2h$ denotes the distance of the source from its image made by the reflecting plane and Z_{A0} means the acoustic driving point impedance for a point source when there is no reflecting plane (free field). Thus, we can express the driving point impedance when the reflecting plane is available as

$$z_{AR} = \frac{p(S)}{Qe^{j\omega t}} = Z_{A0} + j\omega\rho\frac{e^{-2jkh}}{4\pi 2h}.$$

If we take the real part, we have

$$R_R = \frac{\rho c k^2}{4\pi} + \omega\rho\frac{\sin(2kh)}{4\pi 2h} = \frac{\rho c k^2}{4\pi} + \rho c k^2\frac{\sin(2kh)}{4\pi 2kh}$$

$$= \frac{\rho c k^2}{4\pi}\left[1 + \frac{\sin(2kh)}{2kh}\right] = R_0\left[1 + \frac{\sin(2kh)}{2kh}\right] \quad \text{(Pa s m}^{-3}\text{)},$$

where R_0 denotes the real part of the driving impedance of a point source in a free field. Consequently, the power output becomes

$$P = \frac{1}{2}R_R Q^2 = \frac{\rho c k^2 Q^2}{8\pi}\left[1 + \frac{\sin(2kh)}{2kh}\right] = P_0\left[1 + \frac{\sin(2kh)}{2kh}\right] \quad \text{(W)},$$

where P_0 denotes the sound power output of a point source in a free field.

Figure 6.1.6 illustrates the variation of the power output of a point source over a reflecting plane. We see that the power output may decrease from the output obtained in a free field depending on the distance from the reflecting plane. The power output is a basic criterion for selecting the location of a sound source, since the power characteristics do not depend on the sound receiving position. The driving point impedance describes the attribute due to source environment from the viewpoint of power injection into the system to which the sound source is supplied.

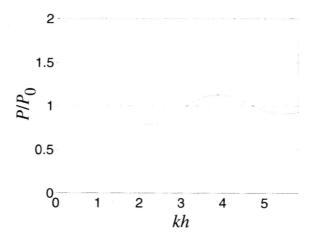

Figure 6.1.6   Sound power output of a point source over reflecting plane.

6.2   Transfer Functions for Acoustic Pipes

6.2.1   An Open Acoustic Pipe with Pressure Driving Source

The solutions of the 1-D plane wave equations provide fundamental models for acoustic transfer functions (TFs). An idealized acoustic pipe is a fundamental model for musical wind instruments such as flutes and clarinets (Rossing and Fletcher, 1995). The input (driving point) and transfer impedances are basic notions for acoustic transfer function analysis and modeling. We will use the terminology of the transfer function (TF) for both the transfer and input impedances in this text.

We describe the wave solution as a complex function, since the complex amplitude (phasor) conveys information on the magnitude and phase. Acoustic quantities such as sound pressure are real numbers. We introduce negative frequency components as complex conjugate quantities in order to reconstruct real functions from the complex wave solutions.

The solution of the wave equation for a 1-D acoustic pipe is obtained in a closed form. Figure 6.2.1 shows an example of an acoustic pipe of length L (m) with both ends open. Wave equation problems are formulated with boundary conditions and driving source conditions. A typical boundary condition for a 1-D acoustic model such as for musical instruments is open end condition. There are two kinds of fundamental source conditions, however, as well as voltage and current sources in electrical circuit theory.

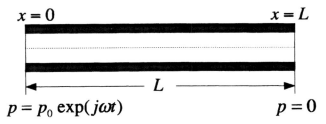

Figure 6.2.1 An acoustic open pipe.

Suppose that a sinusoidal sound pressure driving source is located at one end at $x = 0$. The boundary conditions for the sound pressure waves at both open ends are respectively

$$p = p_0 \exp(j\omega t) \quad \text{(Pa)} \qquad \text{at } x = 0$$

and

$$p = 0 \quad \text{(Pa)} \qquad \text{at } x = L.$$

Substituting these boundary conditions into the solution of the 1-D wave equation given by

$$p = j\omega\rho(A \cos kx + B \sin kx)\exp(j\omega t) \quad \text{(Pa)},$$

we get the following sound pressure distribution in the pipe:

$$p = p_0 \frac{\sin k(L - x)}{\sin kL} \exp(j\omega t) \quad \text{(Pa)},$$

where ρ (kg m^{-3}) denotes the volume density of the medium in the pipe, and k is the wavenumber (m^{-1}).

This solution is derived as follows. The sound pressure satisfies the boundary conditions at both ends

$$p_0 = j\omega\rho A \quad \text{and} \quad 0 = j\omega\rho(A\cos kL + B\sin kL).$$

Thus we can determine the coefficients A and B as

$$A = \frac{p_0}{j\omega\rho} \quad \text{and} \quad B = \frac{-p_0\cos kL}{j\omega\rho\sin kL}.$$

Consequently, we can get the solution for the particular boundary conditions stated above as

$$p = j\omega\rho(A\cos kx + B\sin kx)\exp(j\omega t)$$

$$= j\omega\rho\left(\frac{p_0}{j\omega\rho}\cos kx - \frac{p_0}{j\omega\rho}\frac{\cos kL}{\sin kL}\sin kx\right)\exp(j\omega t)$$

$$= p_0\left(\frac{\sin kL\cos kx - \cos kL\sin kx}{\sin kL}\right)\exp(j\omega t)$$

$$= p_0\frac{\sin k(L-x)}{\sin kL}\exp(j\omega t).$$

The resonance frequencies, which are singularities (where the denominator becomes zero) of the pressure distribution function, are located at

$$k_r = \frac{m\pi}{L} \quad m = 1, 2, 3 \ldots.$$

An open acoustic pipe with a sound pressure source provides a basic model of a flute. The resonance frequencies are arranged in the harmonic structure. Samples of sound pressure distribution patterns (wavemodes) in the pipe and corresponding resonance frequencies are illustrated in Fig. 6.2.2. The loops and nodes that regularly divide the length of the pipe make the typical patterns of wavemodes.

6.2.2 An Open Acoustic Pipe with Velocity Driving Source

Let us consider the other fundamental source conditions in an open acoustic pipe. Suppose that a sinusoidal velocity driving source is located at one end at $x = 0$ instead of a sound pressure source. The boundary conditions at both ends are given by

$$v = v_0\exp(j\omega t) \quad (\text{m s}^{-1}) \quad \text{at } x = 0$$

and

$$p = 0 \quad (\text{m s}^{-1}) \quad \text{at } x = L,$$

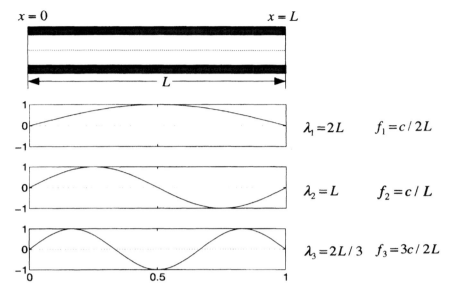

Figure 6.2.2   Examples of pressure wavemodes in an open pipe.

respectively. Substituting these boundary conditions into the solution of the wave equation,

$$p = j\omega\rho(A\cos kx + B\sin kx)\exp(j\omega t),$$

we get the sound pressure distribution in the pipe as well as the solution for the pressure driving source condition. That is, we can determine the coefficients A and B as

$$A = -\frac{\sin kL}{\cos kL}B \qquad \text{and} \qquad B = -\frac{v_0}{k},$$

where

$$j\omega\rho v|_{x=0} = -\frac{\partial p}{\partial x}\bigg|_{x=0} = -j\omega\rho Bke^{j\omega t} = j\omega\rho v_0 e^{j\omega t}$$

and

$$A\cos kL + B\sin kL = 0.$$

Therefore, we have

$$p = j\omega\rho\left(\frac{\sin kL}{\cos kL}\frac{v_0}{k}\cos kx - \frac{v_0}{k}\sin kx\right)$$

$$= j\rho c v_0 \frac{\sin k(L-x)}{\cos kL}\exp(j\omega t) \quad \text{(Pa)},$$

where c (m s^{-1}) is the sound speed.

The resonance frequencies are located at

$$k_r = \frac{(m + \frac{1}{2})\pi}{L} \qquad m = 0, 1, 2, 3, \ldots.$$

We can see that the resonance frequencies are changed by the driving source condition, if we recall that the resonance frequencies under a pressure source are given by

$$k_r = \frac{m\pi}{L} \qquad m = 1, 2, 3 \ldots.$$

This velocity-excited open pipe is a model of a clarinet. The first resonance frequency is 1 octave lower than that of the sound pressure driving condition used for a flute, and the second resonance frequency is 1.5 octave higher than the fundamental frequency.

6.2.3 Driving Point and Transfer Impedances

The transfer function (TF) of an acoustic system is defined as the impedance between the source and receiving positions (x_s, x_r). The particle velocity is given by

$$-\frac{\partial p}{\partial x} = \rho \frac{\partial v}{\partial t} = j\omega \rho v.$$

The velocity at the source position at $x = 0$ in an open pipe with a sound pressure source is

$$v(x = 0) = -\frac{1}{j\omega\rho} \frac{\partial p}{\partial x}\bigg|_{x=0} = -j\frac{p_0}{\rho c} \frac{\cos k(L-x)}{\sin kL} \exp(j\omega t)\bigg|_{x=0}$$

$$= -j\frac{p_0}{\rho c} \frac{\cos kL}{\sin kL} \exp(j\omega t) \quad (\mathrm{m\,s^{-1}}).$$

We have already introduced the acoustic driving point impedance in Section 6.1.11, and we can define the acoustic transfer impedance (or transfer function, TF) between the source and a receiving position as

$$H(\omega, x_s, x_r) \equiv \frac{\text{Sound pressure at the receiving position}}{\text{Source volume velocity}}$$

$$= \frac{p(x_r = x)}{Sv(x_s = 0)} = \frac{p_0}{Sv(x_s = 0)} \frac{\sin k(L-x)}{\sin kL} e^{j\omega t}$$

$$= \frac{j\rho c}{S} \frac{\sin k(L-x)}{\cos kL} \quad (\mathrm{Pa\,s\,m^{-3}}),$$

where we take the source position at $x_s = x = 0$ and the receiving position at $x_r = x$, and S $(\mathrm{m^2})$ denotes the cross section of the pipe.

The zeros of the TF are defined as the complex roots of the TF, while the poles are singularities. The poles and zeros are given by

$$k_p = \frac{(n + \frac{1}{2})\pi}{L} \qquad \text{for poles}$$

and

$$k_z = \frac{n\pi}{L - x} \qquad \text{for zeros},$$

respectively. The poles are uniformly distributed on the frequency axis, while the locations of zeros depend on the receiving and source positions.

When the receiving position x is set at the source position, we can define the acoustic input impedance (or driving point impedance) for a special case of the transfer function. The input impedance for the open acoustic pipe where the sound pressure source is located at one end is

$$Z(\omega, x_s) \equiv \frac{\text{Sound pressure at the source position } x_s}{\text{Source volume velocity}}$$

$$= \frac{p(x_r = 0)}{Sv(x_s = 0)} = \frac{j\rho c}{S} \frac{\sin kL}{\cos kL} = \frac{j\rho c}{S} \tan(kL) \quad (\text{Pa s m}^{-3}).$$

The resonance frequencies for the sound pressure source, which are given by

$$k_r = \frac{m\pi}{L} \qquad m = 1, 2, 3 \ldots,$$

correspond to the zeros of the input impedance.

The transfer and input impedances with the sound pressure source are the same as those for the pipe with the velocity driving source. The sound pressure at a receiving position is given by

$$p = j\rho c v_0 \frac{\sin k(L - x)}{\cos kL} \exp(j\omega t) \quad (\text{Pa})$$

when the velocity driving source is located at $x_s = 0$. The transfer and the input impedances become

$$H(\omega) = \frac{j\rho c}{S} \frac{\sin k(L - x)}{\cos kL} \quad (\text{Pa s m}^{-3})$$

and

$$Z(\omega) = \frac{j\rho c}{S} \tan(kL) \quad (\text{Pa s m}^{-3}),$$

respectively. Therefore, the transfer function, which is used for both impedances in this text, is an intrinsic property of acoustic systems.

The resonance frequencies under the velocity source condition, however, correspond to the poles of the input impedance.

Figure 6.2.3 shows examples of magnitude characteristics for the input and transfer impedances, for both of which we will use the term the transfer function in this text. The poles and zeros of the input impedance are interlaced as shown in Fig. 6.2.3a. The poles are not changed according to the receiving positions, but the zeros are shifted into the high frequencies as the distance between the source and receiving positions increases (Lyon, 1984).

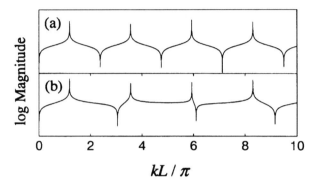

Figure 6.2.3   Image of log magnitudes of input (a) and transfer (b) impedances in an open pipe.

As we have already described in Section 6.1.12, the real part of the driving point impedance represents the power injected into a system from a driving source. There are remarkable differences between the acoustic pipe and a point source over a reflecting plane discussed in Section 6.1.12. In the open pipe discussed in this section, we have no real parts in the driving point impedance. This is because we assume no damping in the system. Such a system without any loss corresponds to an *LC* circuit. And the driving point impedance for the pipe has poles and zeros that we did not see in Section 6.1.12. These are due to a series of an infinite number of reflections from both ends, while the reflecting plane discussed in Section 6.1.12 produces a single reflection.

6.3   Eigenfrequencies and Eigenfunctions

6.3.1   Wave Equations and Eigenvalues

The wave equation is written

$$\nabla^2 u = \frac{1}{c^2} \frac{\partial^2 u}{\partial t^2}.$$

For periodic solutions under sinusoidal source condition

$$u(\mathbf{x}, t) = u(\mathbf{x})e^{j\omega t}$$

the wave equation can be rewritten as

$$\nabla^2 u + k^2 u = 0,$$

where k is the wavenumber and $k = \omega/c$.

The general form of the wave equation for the periodic waves is expressed as (Mathews and Waker, 1973)

$$Lu(\mathbf{x}) = \lambda u(\mathbf{x}),$$

where L is a linear differential operator and λ is its eigenvalue. The solution for this equation is called the eigenfunction corresponding to the eigenvalue λ. This is similar to the linear transform by a matrix,

$$A\mathbf{x} = \lambda \mathbf{x},$$

where A denotes the matrix, λ is the eigenvalue, and $\mathbf{x}$ denotes the eigenvector corresponding to the eigenvalue.

A linear operator L is called Hermitian (Mathews and Waker, 1973) if

$$\int_\Omega u^*(\mathbf{x})Lv(\mathbf{x})\,d\mathbf{x} = \left[\int_\Omega v^*(\mathbf{x})Lu(\mathbf{x})\,d\mathbf{x}\right]^*,$$

where Ω is a specified region for the operator where suitable boundary conditions are imposed. Both u and v are arbitrary functions which obey the boundary conditions.

6.3.2 Eigenfunctions and Orthogonality

The eigenvalues of Hermitian operators are real and the eigenfunctions corresponding to different eigenvalues are orthogonal in the specified region (Mathews and Waker, 1973). Suppose the linear operator L is Hermitian. We have

$$Lu_i(\mathbf{x}) = \lambda_i u_i(\mathbf{x}),$$

where λ_i denotes the ith eigenvalue and $u_i(\mathbf{x})$ is the eigenfunction for the ith eigenvalue. We take another pair of the eigenvalue and eigenfunction such as

$$Lu_j(\mathbf{x}) = \lambda_j u_j(\mathbf{x}).$$

Then we get the relations in the specified region Ω as

$$\int_\Omega u_j^*(\mathbf{x})Lu_i(\mathbf{x})\,d\mathbf{x} = \lambda_i \int_\Omega u_j^*(\mathbf{x})u_i(\mathbf{x})\,d\mathbf{x}$$

and

$$\int_\Omega u_i^*(\mathbf{x})Lu_j(\mathbf{x})\,d\mathbf{x} = \lambda_j \int_\Omega u_i^*(\mathbf{x})u_j(\mathbf{x})\,d\mathbf{x}.$$

Since L is Hermitian,

$$\int_\Omega u_j^*(\mathbf{x})Lu_i(\mathbf{x})\,d\mathbf{x} = \lambda_i \int_\Omega u_j^*(\mathbf{x})u_i(\mathbf{x})\,d\mathbf{x} = \left[\int_\Omega u_i^*(\mathbf{x})Lu_j(\mathbf{x})\,d\mathbf{x}\right]^*$$

$$= \left[\lambda_j \int_\Omega u_i^*(\mathbf{x})u_j(\mathbf{x})\,d\mathbf{x}\right]^* = \lambda_j^* \int_\Omega u_i(\mathbf{x})u_j^*(\mathbf{x})\,d\mathbf{x}.$$

Thus we get the orthogonality for the eigenfunctions as

$$\lambda_i \int_\Omega u_j^*(\mathbf{x})u_i(\mathbf{x})\,d\mathbf{x} = \lambda_j^* \int_\Omega u_i(\mathbf{x})u_j^*(\mathbf{x})\,d\mathbf{x}$$

or

$$(\lambda_i - \lambda_j^*) \int_\Omega u_j^*(\mathbf{x})u_i(\mathbf{x})\,d\mathbf{x} = 0.$$

Consequently,

$$\int_\Omega u_j^*(\mathbf{x})u_i(\mathbf{x})\,d\mathbf{x} = 0,$$

when $\lambda_i \neq \lambda_j$ and when $i = j$, $\lambda_i = \lambda_j^*$. Therefore, the eigenvalues are real.

6.3.3 Green's Theorem

Take two vectors $\mathbf{A}$ and $\mathbf{B}$ which are functions of (x, y, z) and are related as

$$\mathbf{A} = \psi\mathbf{B},$$

where ψ is a scalar function of (x, y, z). The divergence of $\mathbf{A}$ is given by

$$\operatorname{div}\mathbf{A} = \operatorname{div}(\psi\mathbf{B}) = \frac{\partial(\psi B_x)}{\partial x} + \frac{\partial(\psi B_y)}{\partial y} + \frac{\partial(\psi B_z)}{\partial z}$$

$$= \psi\left(\frac{\partial B_x}{\partial x} + \frac{\partial B_y}{\partial y} + \frac{\partial B_z}{\partial z}\right) + \frac{\partial\psi}{\partial x}B_x + \frac{\partial\psi}{\partial y}B_y + \frac{\partial\psi}{\partial z}B_z$$

$$= \psi\operatorname{div}\mathbf{B} + \nabla\psi\cdot\mathbf{B}$$

where

$$\operatorname{div}\mathbf{B} = \frac{\partial B_x}{\partial x} + \frac{\partial B_y}{\partial y} + \frac{\partial B_z}{\partial z}$$

and

$$\nabla\psi = \frac{\partial\psi}{\partial x}\mathbf{i} + \frac{\partial\psi}{\partial y}\mathbf{j} + \frac{\partial\psi}{\partial z}\mathbf{k}.$$

Here $(\mathbf{i}, \mathbf{j}, \mathbf{k})$ denote the orthogonal unit vectors of the (x, y, z) space.

The divergence can be defined through an integral formula such as

$$\int_S \mathbf{A} \cdot \mathbf{dS} = \int_V \operatorname{div} \mathbf{A} \, dv,$$

where S denotes a closed surface in the space where the vector $\mathbf{A}$ is defined, and V is the volume of the area surrounded by the surface S. Introducing the relation $\mathbf{A} = \psi\mathbf{B}$ into the definition of divergence, we get

$$\int_S \mathbf{A} \cdot \mathbf{dS} = \int_S \psi\mathbf{B} \cdot \mathbf{dS} = \int_V \operatorname{div} \psi\mathbf{B} \, dv.$$

And again introducing

$$\operatorname{div} \psi\mathbf{B} = \psi \operatorname{div} \mathbf{B} + \nabla\psi \cdot \mathbf{B}$$

we get

$$\int_S \psi\mathbf{B} \cdot \mathbf{dS} = \int_V \operatorname{div} \psi\mathbf{B} \, dv = \int_V (\psi \operatorname{div} \mathbf{B} + \nabla\psi \cdot \mathbf{B}) \, dv.$$

Let us redefine the vector $\mathbf{B}$ using a new scalar variable ϕ as

$$\mathbf{B} = \operatorname{grad} \phi = \nabla\phi.$$

The divergence of $\mathbf{B}$ can be rewritten as

$$\operatorname{div} \mathbf{B} = \nabla \cdot \mathbf{B} = \nabla \cdot \nabla\phi = \nabla^2\phi.$$

Therefore, we have

$$\int_S \psi\mathbf{B} \cdot \mathbf{dS} = \int_S \psi\nabla\phi \cdot \mathbf{dS}$$

and

$$\int_V (\psi \operatorname{div} \mathbf{B} + \nabla\psi \cdot \mathbf{B}) \, dv = \int_V (\psi\nabla^2\phi + \nabla\psi \cdot \nabla\phi) \, dv.$$

Connecting these two equations by

$$\int_S \psi\mathbf{B} \cdot \mathbf{dS} = \int_V (\psi \operatorname{div} \mathbf{B} + \nabla\psi \cdot \mathbf{B}) \, dv,$$

we get

$$\int_S \psi\nabla\phi \cdot \mathbf{dS} = \int_V (\psi\nabla^2\phi + \nabla\psi \cdot \nabla\phi) \, dv.$$

If we exchange the variables, we obtain

$$\int_S \phi\nabla\psi \cdot \mathbf{dS} = \int_V (\phi\nabla^2\psi + \nabla\phi \cdot \nabla\psi) \, dv.$$

The difference between these two equations becomes

$$\int_S (\psi \nabla \phi - \phi \nabla \psi) \cdot \mathbf{dS} = \int_V (\psi \nabla^2 \phi - \phi \nabla^2 \psi) \, dv$$

or

$$\int_S \left(\psi \frac{\partial \phi}{\partial n} - \phi \frac{\partial \psi}{\partial n} \right) dS = \int_V (\psi \nabla^2 \phi - \phi \nabla^2 \psi) \, dv.$$

This is known as Green's theorem. Here $(\partial/\partial n)$ denotes taking the gradient normal.

6.3.4   Vibration of a Circular Membrane

Examples of eigenfunctions can be seen in the vibration of a circular membrane (Rossing and Fletcher, 1994; Mathews and Waker, 1973). The wave equation in polar coordinates for a circular membrane of radius a is written as follows. The wave equation

$$\nabla^2 u = \frac{1}{c^2} \frac{\partial^2 u}{\partial t^2}$$

is rewritten as

$$\frac{\partial^2 u}{\partial r^2} + \frac{1}{r} \frac{\partial u}{\partial r} + \frac{1}{r^2} \frac{\partial^2 u}{\partial \theta^2} = \frac{1}{c^2} \frac{\partial^2 u}{\partial t^2}$$

in polar coordinates, where u denotes the displacement of the membrane (m), and

$$\nabla^2 u = \frac{1}{r} \frac{\partial}{\partial r} \left(r \frac{\partial u}{\partial r} \right) + \frac{1}{r^2} \frac{\partial^2 u}{\partial \theta^2} = \frac{\partial^2 u}{\partial r^2} + \frac{1}{r} \frac{\partial u}{\partial r} + \frac{1}{r^2} \frac{\partial^2 u}{\partial \theta^2}.$$

If we set the solution of this wave equation as

$$u = U(r, \theta) e^{j\omega t} \quad \text{(m)},$$

we get the equation

$$\frac{\partial^2 U}{\partial r^2} + \frac{1}{r} \frac{\partial U}{\partial r} + \frac{1}{r^2} \frac{\partial^2 U}{\partial \theta^2} + k^2 U = 0,$$

where $k = \omega/c$ denotes the wavenumber, and c is the wave propagation speed $(\mathrm{m\,s^{-1}})$. Again setting the solution in the form

$$U \equiv R(r)\vartheta(\theta) \quad \text{(m)}$$

in order to separate the variables, we get the equation

$$\vartheta \frac{d^2R}{dr^2} + \vartheta \frac{1}{r}\frac{dR}{dr} + R\frac{1}{r^2}\frac{d^2\vartheta}{d\theta^2} + k^2 R\vartheta = 0.$$

If we divide both sides by $R\vartheta$, then the above equation becomes

$$\frac{1}{R}\frac{d^2R}{dr^2} + \frac{1}{Rr}\frac{dR}{dr} + \frac{1}{\vartheta r^2}\frac{d^2\vartheta}{d\theta^2} + k^2 = 0.$$

Multiplying by r^2, we get

$$\frac{r^2}{R}\frac{d^2R}{dr^2} + \frac{r}{R}\frac{dR}{dr} + \frac{1}{\vartheta}\frac{d^2\vartheta}{d\theta^2} + r^2 k^2 = 0,$$

or equivalently

$$\frac{r^2}{R}\frac{d^2R}{dr^2} + \frac{r}{R}\frac{dR}{dr} + r^2 k^2 = -\frac{1}{\vartheta}\frac{d^2\vartheta}{d\theta^2}.$$

Consequently, we can separate the variables so that the right-hand term depends only on θ, while the rest depend only on r. Therefore,

$$\frac{1}{\vartheta}\frac{d^2\vartheta}{d\theta^2} = \text{const} \equiv -m^2$$

or

$$\frac{d^2\vartheta}{d\theta} = -m^2\vartheta,$$

where m^2 is called the separation constant. Similarly, the equation for R becomes

$$\frac{r^2}{R}\frac{d^2R}{dr^2} + \frac{r}{R}\frac{dR}{dr} + r^2 k^2 = m^2,$$

or equivalently

$$\left(\frac{d^2}{dr^2} + \frac{1}{r}\frac{d}{dr} - \frac{m^2}{r^2} \right) R = -\frac{\omega^2}{c^2}R.$$

The solution to the first equation is

$$\vartheta(\theta) = Ae^{\pm jm\theta},$$

where m corresponds to the eigenvalue as we will see later. The second equation can be rewritten as

$$\frac{d^2R}{dx^2} + \frac{1}{x}\frac{dR}{dx} + \left(1 - \frac{m^2}{x^2}\right)R = 0,$$

where $x = kr = \omega r/c$. This is called Bessel's differential equation. The solutions to Bessel's equation are Bessel functions of order m. Some Bessel functions $J_m(x)$ are illustrated in Fig. 6.3.1.

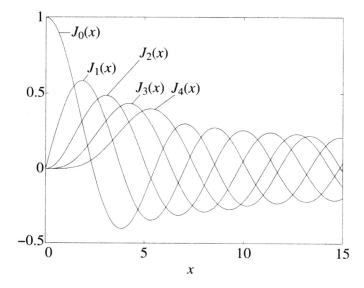

Figure 6.3.1   Examples of Bessel functions $J_m(x)$.

The mth-order wave functions for the vibration of a circular membrane are composed of

$$u_{\text{even}} = R_m(r)\cos m\theta e^{j\omega t} = J_m(kr)\cos m\theta e^{j\omega t}$$

and

$$u_{\text{odd}} = R_m(r)\sin m\theta e^{j\omega t} = J_m(kr)\sin m\theta e^{j\omega t}$$

where m must be an integer, since the vibration solution must be single valued and it is a periodic function of θ with period 2π. Each of the Bessel functions becomes zero at several values of x. The boundary condition imposed on the membrane vibration is $R = 0$ at $r = a$. Thus, we require

$$J_m(ka) = 0.$$

Suppose that the nth root of the mth-order Bessel function is

$$k_{mn}a \equiv u_{mn}.$$

The eigenfrequency for the (m, n) eigenfunction (wavemode) is obtained as

$$k_{mn} = \frac{u_{mn}}{a} = \frac{2\pi f_{mn}}{c},$$

or

$$f_{mn} = c\frac{u_{mn}}{2\pi a} \quad (\text{Hz}).$$

Several roots of the Bessel functions are given in Table 6.3.1.

Figure 6.3.2 shows samples of the patterns of eigenfunctions for the circular membrane. There are two eigenfunctions, which are the even and odd functions u_{even} and u_{odd}, with respect to the angular variable θ belonging to the same eigenfrequencies except for the first eigenfrequency. This is called degeneracy of eigenfrequencies and is due to the symmetric property of the systems. The orthogonality of Bessel functions which belong to different eigenfrequencies is satisfied when

$$\int_0^a \psi_{mn}\psi_{m'n'}r\,dr = \begin{cases} 0 & (n' \neq n, \text{ or } m' \neq m) \\ \pi a^2 \Lambda_{mn} & (n' = n, \text{ and } m' = m) \end{cases}$$

where

$$\psi_{mn} = J_{mn}\left(\frac{u_{mn}}{a}r\right),$$

$$\Lambda_{0_n} = J_1^2(u_{0_n}),$$

and

$$\Lambda_{mn} = \tfrac{1}{2}J_{m-1}^2(u_{mn}) \quad (m > 0).$$

Table 6.3.1 Examples of mth roots of Bessel functions.

			n		
m	1	2	3	4	5
0	2.404 83	5.520 08	8.653 73	11.791 53	14.930 92
1	3.831 71	7.015 59	10.173 47	13.323 69	16.470 63
2	5.135 62	8.417 24	11.619 84	14.795 95	17.959 82
3	6.380 16	9.761 02	13.015 20	16.223 47	19.409 42
4	7.588 34	11.064 71	14.372 54	17.616 97	20.826 93

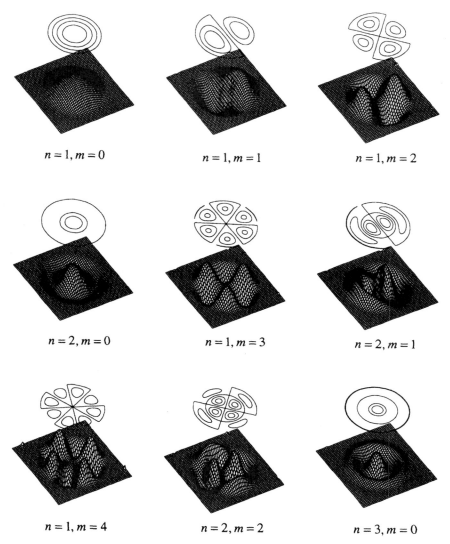

Figure 6.3.2   Examples of eigenfunctions of circular membrane vibrations.

6.4   Distribution of Eigenfrequencies

6.4.1   Degeneracy of Eigenfrequencies

The eigenfrequencies for a 1-D system are distributed uniformly. The eigenfrequencies for 2-D or 3-D systems, however, are distributed irregularly. For a rectangular membrane whose edges are clamped, the mth eigenfrequency is given by

$$\omega_{lm} = c\sqrt{\left(\frac{l\pi}{L_x}\right)^2 + \left(\frac{m\pi}{L_y}\right)^2} \quad (\text{rad s}^{-1}),$$

where L_x and L_y denote the lengths of the sides in meters and c denotes the wave speed (m s^{-1}).

If the ratio between L_x and L_y, which is related to the symmetry of the system, is

$$L_y = \frac{L_x}{\alpha} \quad (\text{m}),$$

where α is an integer, then the eigenfrequency is rewritten as

$$\omega_{lm} = c\sqrt{\left(\frac{l\pi}{L_x}\right)^2 + \left(\frac{\alpha m\pi}{L_x}\right)^2} = c\left(\frac{\pi}{L_x}\right)\sqrt{l^2 + \alpha^2 m^2} \quad (\text{rad s}^{-1}).$$

Degeneracy, $\omega_{lm} = \omega_{pl}$, occurs where $p = \alpha m$.

6.4.2 Poisson Distribution

The occurrence of eigenfrequencies in a frequency interval can be modeled as a Poisson process, since eigenfrequency occurrence in any unit frequency interval is equally likely in a finite interval of interest to us. A Poisson process is a model of random events which occur with equal probability in every small interval Δt. Let us take an interval T divided by M:

$$\Delta t = \frac{T}{M}.$$

Suppose that the average number of events which occur in the interval is νT. The probability of one event occurring in an interval Δt is given by

$$\frac{\nu T}{M} = \nu \Delta t.$$

This can be considered to be much smaller than unity and we can neglect the probability with which we have more than a single event in an interval Δt.

The probability for the case in which the events occur k times in the interval T becomes

$$P(k) = \frac{M!}{k!(M-k)!}(\nu \Delta t)^k (1 - \nu \Delta t)^{M-k},$$

where $(\nu \Delta t)^k (1 - \nu \Delta t)^{M-k}$ denotes the probability with which k intervals have events, $M - k$ intervals have no events, and $M!/[k!(M-k)!]$ is the number of combinations for sampling k intervals from the M intervals. Introducing

$$\Delta t = \frac{T}{M}$$

we have

$$P(k) = \frac{M!}{k!(M-k)!}(v\Delta t)^k(1-v\Delta t)^{M-k}$$

$$= \frac{M!}{k!(M-k)!}\left(v\frac{T}{M}\right)^k\left(1-v\frac{T}{M}\right)^{M-k}$$

$$= \frac{M(M-1)\cdots(M-k+1)}{k!M^k}(vT)^k\left(1-v\frac{T}{M}\right)^{M-k}$$

$$= \frac{\left(1-\dfrac{1}{M}\right)\cdots\left(1-\dfrac{k-1}{M}\right)}{k!}(vT)^k\left(1-v\frac{T}{M}\right)^{M-k}.$$

Taking the limit for M going to infinity, we get

$$P(k) = \frac{(vT)^k}{k!}\exp(-vT),$$

where

$$\left(1-v\frac{T}{M}\right)^{M-k} \to \left[\left(1-\frac{1}{M/vT}\right)^{-M/vT}\right]^{-vT}$$

$$\to \exp(-vT)\qquad(M\to\infty).$$

Here the relation as a limit

$$\left(1-\frac{1}{y}\right)^{-y} = \left(\frac{y-1}{y}\right)^{-y} = \left(\frac{y}{y-1}\right)^{y} = \left(1+\frac{1}{y-1}\right)^{y}$$

$$= \left(1+\frac{1}{y-1}\right)^{y-1}\left(1+\frac{1}{y-1}\right)$$

$$\to \left(1+\frac{1}{y}\right)^{y}\qquad(y\to\text{ large})$$

$$\to e(y\to\infty)$$

is used.

   The probability following a Poisson process is given by

$$P(k) = \frac{(vT)^k}{k!}\exp(-vT).$$

This gives the probability for the case that the random event occurs k times in the time interval T where νT denotes the expected value of the times when the event occurs in the interval and ν is the average of the event occurrence in the unit interval.

The probability density $q(\tau)$ for the spacing τ in the two events which occur successively is obtained by solving,

$$\int_{\tau}^{\infty} q(\tau)\, d\tau = \exp(-\nu T),$$

where the right-hand side of the equation gives the probability of having no events ($k = 0$) within the spacing τ, and thus the left-hand side denotes the probability for the case that the event occurs after τ. Solving this equation,

$$q(\tau) = \nu \exp(-\nu \tau).$$

This is an exponential distribution.

6.4.3 Eigenfrequency Spacing Distribution

The degeneracy possibilities are related to spacing statistics of eigenfrequencies. The spacing of a pair of adjacent eigenfrequencies for a 2-D system is distributed following an exponential distribution as shown in Fig. 6.4.1. Here L_x and L_y are lengths of the sides (m), and c is the propagation speed (m s^{-1}). This exponential distribution of the eigenfrequency spacings means that the distribution of eigenfrequencies in a frequency interval is modeled by a Poisson process.

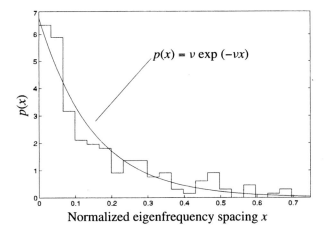

Figure 6.4.1 Eigenfrequency spacing distribution for a 2-D system.

Spacing statistics following the exponential distribution show that degeneracy is quite likely, since the most probable spacing is 0. However, the degeneracy seems quite unlikely in most practical cases such as a reverberation room because of the breaking of the symmetry which is the source of the degeneration (Lyon, 1969; Schroeder, 1989). Figure 6.4.2 shows an example of the spacing statistics for a 2-D stadium field where eigenvalues are calculated by the Finite Element Method. We can see a so-called Wigner distribution where the degeneracy is quite unlikely to be seen.

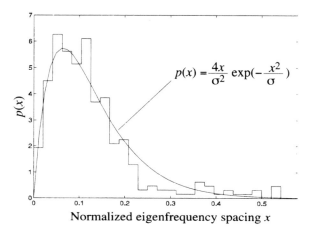

$$p(x) = \frac{4x}{\sigma^2} \exp(-\frac{x^2}{\sigma})$$

Normalized eigenfrequency spacing x

Figure 6.4.2   Eigenfrequency spacing statistics for a 2-D stadium field.

6.4.4   Number of Eigenfrequencies

The modal density is defined by the average number of eigenfrequencies in a unit frequency interval. The number of the eigenfrequencies in a 2-D system below a test angular frequency $\omega = ck$ (rad s^{-1}) is given by (Morse and Bolt, 1944)

$$N(k) \cong \frac{S}{4\pi}k^2 + \frac{L}{4\pi}k \qquad \text{for rectangular boundaries}$$

and

$$N(k) \cong \frac{S}{4\pi}k^2 - \frac{L}{4\pi}k \qquad \text{for nonrectangular boundaries,}$$

where S (m^2) denotes the area inside the boundary and L (m) is the length of the circumference.

Let us take the example of a rectangular case. The eigenvalues of the wavenumber are given by

$$k_{lm} = \sqrt{\left(\frac{l\pi}{L_x}\right)^2 + \left(\frac{m\pi}{L_y}\right)^2}.$$

These eigenvalues are distributed on the lattice points on the rectangular two-dimensional wavenumber space as shown in Fig. 6.4.3. The number of eigenvalues which are located below wavenumber k can be estimated as follows.

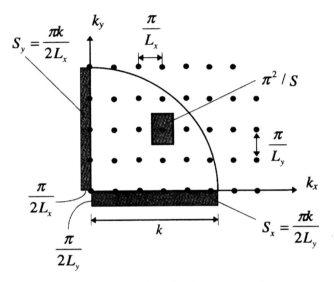

Figure 6.4.3 Eigenvalues in the wavenumber space.

The total area of the quarter circle with radius k and the areas along the k_x and k_y axes as shown in Fig. 6.4.3 is given by

$$D(k) = \frac{\pi k^2}{4} + \frac{\pi k}{2L_x} + \frac{\pi k}{2L_y} = \frac{\pi k^2}{4} + \frac{\pi k}{4}\frac{L}{S},$$

where $S = L_x L_y$ and $L = 2(L_x + L_y)$. The number of lattice points included in this area gives the number of the eigenvalues,

$$N(k) = \frac{D(k)}{\pi^2/S} = \frac{\pi k^2/4}{\pi^2/S} + \frac{(\pi k/4)(L/S)}{\pi^2/S} = \frac{Sk^2}{4\pi} + \frac{Lk}{4\pi}.$$

This can be extended into estimation of the number of eigenvalues for nonrectangular cases by subtracting the number of the lattice points on the k_x and k_y axes. Thus we have

$$N(k) = \frac{Sk^2}{4\pi} + \frac{Lk}{4\pi} - \left(\frac{kL_x}{\pi} + \frac{kL_y}{\pi}\right) = \frac{Sk^2}{4\pi} - \frac{Lk}{4\pi}.$$

Figure 6.4.4 illustrates examples of the number of eigenfrequencies for 2-D systems. The number of eigenfrequencies of the rectangular system can be estimated by the sum of the axial and tangential wavemodes. Axial wavemodes have the eigenfrequencies

$$\omega_{lm} = c\sqrt{\left(\frac{l\pi}{L_x}\right)^2 + \left(\frac{m\pi}{L_y}\right)^2}$$

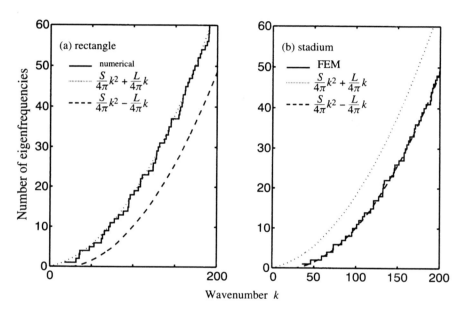

Figure 6.4.4 Accumulated number of eigenfrequencies (a) rectangular (b) stadium.

where l or m is 0, while l and m of tangential wavemodes are not 0. The accumulated number of eigenfrequencies for the nonrectangular boundary can be estimated by the number of tangential wavemodes. The densities of the eigenfrequencies for 2-D systems are given by the derivatives of $N(k)$. These approximation formulae can be extended into 3-D cases by (Morse and Bolt, 1944)

$$N(k) = \frac{V}{6\pi^2}k^3 + \frac{S}{16\pi}k^2 + \frac{L}{16\pi}k \qquad \text{for rectangular}$$

and

$$N(k) = \frac{V}{6\pi^2}k^3 - \frac{S}{16\pi}k^2 + \frac{L}{16\pi}k \qquad \text{for nonrectangular,}$$

where V denotes the volume (m³).

6.5 Green's Function

6.5.1 Orthogonal Function Expansion

As was described in Section 6.4, the eigenfunctions for an Hermitian operator are orthogonal. Consider an expansion of the form

$$-f(\mathbf{x}) = \sum_n c_n u_n(\mathbf{x}).$$

When $u_n(\mathbf{x})$ is orthonormal in a specific region, $u_n(\mathbf{x})$ satisfies the orthonormal relationships

$$\int_\Omega u_n^*(\mathbf{x}) u_n(\mathbf{x})\, d\mathbf{x} = 1$$

and

$$\int_\Omega u_n^*(\mathbf{x}) u_m(\mathbf{x})\, d\mathbf{x} = 0$$

for $n \neq m$. Following this relationship the coefficient c_n can be determined as

$$-\int_\Omega u_n^*(\mathbf{x}) f(\mathbf{x})\, d\mathbf{x} = \sum_m c_m \int_\Omega u_n^*(\mathbf{x}) u_m(\mathbf{x})\, d\mathbf{x} = c_n.$$

If we substitute this result back into the original series, we get

$$-f(\mathbf{x}) = -\sum_n \left(\int_\Omega u_n^*(\mathbf{x}') f(\mathbf{x}')\, d\mathbf{x}' \right) u_n(\mathbf{x})$$

$$= -\int_\Omega f(\mathbf{x}')\, d\mathbf{x}' \sum_n u_n^*(\mathbf{x}') u_n(\mathbf{x})$$

and consequently we have

$$\sum_n u_n^*(\mathbf{x}') u_n(\mathbf{x}) = \delta(\mathbf{x} - \mathbf{x}').$$

The relation above between the orthonormal functions and the delta function is the so-called completeness relation.

We assume that our operator is Hermitian so that the eigenfunctions corresponding to different eigenvalues are orthogonal (orthonormal) and complete. Suppose that we have the inhomogeneous equation

$$Lu(\mathbf{x}) - \lambda u(\mathbf{x}) = -f(\mathbf{x})$$

over a domain Ω where λ is a given constant. The right-hand term corresponds to the spatial distribution of external forces having $e^{j\omega t}$ time dependency. This is because the wave equation with an external force

$$\frac{\partial^2 u}{\partial t^2} = c^2 \nabla^2 u + F$$

can be rewritten as

$$-\omega^2 u(\mathbf{x}) e^{j\omega t} = c^2 \nabla^2 u(\mathbf{x}) e^{j\omega t} + F(\mathbf{x}) e^{j\omega t},$$

or equivalently

$$\nabla^2 u(\mathbf{x}) + k^2 u(\mathbf{x}) = \nabla^2 u(\mathbf{x}) - \lambda u(\mathbf{x}) = -f(\mathbf{x}),$$

where $k^2 = -\lambda$. We impose suitable boundary conditions on the equation.

Both $u(\mathbf{x})$ and $f(\mathbf{x})$ can be expanded in eigenfunctions of the operator L with the boundary conditions

$$u(\mathbf{x}) = \sum_n c_n u_n(\mathbf{x}) \qquad \text{and} \qquad -f(\mathbf{x}) = \sum_n d_n u_n(\mathbf{x}).$$

Here, the nth eigenfunction $u_n(\mathbf{x})$ satisfies the homogeneous equation

$$L u_n(\mathbf{x}) = \lambda_n u_n(\mathbf{x})$$

and the imposed boundary conditions with the nth eigenvalue λ_n. If we introduce the equations above for the original inhomogeneous equation,

$$L u(\mathbf{x}) - \lambda u(\mathbf{x}) = -f(\mathbf{x}),$$

we get a new equation given by

$$\sum_n c_n L u_n(\mathbf{x}) - \lambda \sum_n c_n u_n(\mathbf{x}) = \sum_n d_n u_n(\mathbf{x}),$$

or equivalently

$$\sum_n c_n (\lambda_n - \lambda) u_n(\mathbf{x}) = \sum_n d_n u_n(\mathbf{x}).$$

Consequently the coefficient c_n is given by

$$c_n = \frac{d_n}{\lambda_n - \lambda} = \frac{-u_n \cdot f}{\lambda_n - \lambda},$$

where

$$u_n \cdot f = \int_\Omega u_n^*(\mathbf{x}') f(\mathbf{x}') \, d\mathbf{x}'.$$

Here the eigenfunctions are normalized as

$$\int_\Omega u_i^*(\mathbf{x}) u_j(\mathbf{x}) \, d\mathbf{x} = \delta_{ij}.$$

Therefore,

$$u(\mathbf{x}) = -\sum_n \frac{u_n(\mathbf{x})u_n \cdot f}{\lambda_n - \lambda} = \sum_n \frac{u_n(\mathbf{x})}{\lambda - \lambda_n} \int_\Omega u_n^*(\mathbf{x}')f(\mathbf{x}')\,d\mathbf{x}'$$

$$= \int_\Omega \sum_n \frac{u_n(\mathbf{x})u_n^*(\mathbf{x}')}{\lambda - \lambda_n} f(\mathbf{x}')\,d\mathbf{x}' = \int_\Omega G(\mathbf{x},\mathbf{x}')f(\mathbf{x}')\,d\mathbf{x}'$$

where $G(\mathbf{x},\mathbf{x}')$ is called a Green's function and is given by

$$G(\mathbf{x},\mathbf{x}') = \sum_n \frac{u_n(\mathbf{x})u_n^*(\mathbf{x}')}{\lambda - \lambda_n}.$$

When the external force is a unit point source located at position $\mathbf{x}_0$ as

$$f(\mathbf{x}) = \delta(\mathbf{x} - \mathbf{x}_0),$$

the solution $u(\mathbf{x})$ for the inhomogeneous equation

$$Lu(\mathbf{x}) - \lambda u(\mathbf{x}) = -\delta(\mathbf{x} - \mathbf{x}_0),$$

can be written as

$$u(\mathbf{x}) = \int_\Omega G(\mathbf{x},\mathbf{x}')\delta(\mathbf{x}' - \mathbf{x}_0)\,d\mathbf{x}' = G(\mathbf{x},\mathbf{x}_0).$$

Therefore, the Green's function is the solution of

$$LG(\mathbf{x},\mathbf{x}') - \lambda G(\mathbf{x},\mathbf{x}') = -\delta(\mathbf{x} - \mathbf{x}')$$

under the appropriate boundary conditions.

We can see a reciprocal relation

$$G(\mathbf{x},\mathbf{x}') = \sum_n \frac{u_n(\mathbf{x})u_n^*(\mathbf{x}')}{\lambda - \lambda_n} = \left[\sum_n \frac{u_n(\mathbf{x}')u_n^*(\mathbf{x})}{\lambda - \lambda_n}\right]^* = [G(\mathbf{x}',\mathbf{x})]^*$$

between the observation and the source positions, where the eigenvalues are real because of a Hermitian operator. We will take this Green's function as a general model for the transfer function in acoustic systems, although Green's functions do not have the dimension of impedance. This is because Green's functions can be interpreted as the Fourier transform of the impulse response of an acoustic system which is excited by a point source, as we will see in Section 6.5.4.

6.5.2 Orthogonal Expansion of Green's Function for a 1-D Wave Equation

An example of Green's functions is illustrated using the 1-D homogeneous wave equation in the interval 0 to L. Assuming a time dependency of $e^{j\omega t}$

for the solution, the wave equation becomes

$$\frac{d^2u}{dx} + k^2u = 0,$$

where $k = \omega/c$ and c (m s^{-1}) denotes the sound speed. The equation is rewritten as

$$\frac{d^2u}{dx} = -k^2u = \lambda u,$$

where $\lambda = -k^2$.

Here let us impose simpler boundary conditions, $u(0) = u(L) = 0$ (m) than those required for acoustic pipes which are described in Section 6.2. These boundary conditions are generally imposed for the vibrations of a finite string both of whose ends are fixed. This is a fundamental model of the strings of a piano or a guitar. The boundary conditions above can be extended to 3-D room acoustics problems, when the surrounding walls are hard.

By imposing the boundary conditions above, the eigenvalues and normalized eigenfunctions are

$$\lambda_n = -k_n^2 = -\left(\frac{n\pi}{L}\right)^2 \qquad \text{with } n = 1, 2, 3, \ldots$$

and

$$u_n = \sqrt{\frac{2}{L}} \sin \frac{n\pi x}{L}.$$

This can be confirmed as follows. Since $u = 0$ when $x = 0$, then we set a form of an eigenfunction as $u = A \sin kx$, where A is an arbitrary coefficient. This form satisfies the equation

$$\frac{d^2u}{dx} = -k^2u = \lambda u$$

since

$$\frac{du}{dx} = Ak \cos kx$$

and

$$\frac{d^2u}{dx^2} = -Ak^2 \sin kx = -k^2u.$$

But this function must satisfy $u = 0$ at $x = L$ because of the boundary conditions imposed on the equation. Thus the eigenvalues are obtained as

$$k = \frac{n\pi}{L} \qquad n = 1, 2, 3, \ldots$$

by solving

$$\sin kL = 0.$$

By taking the normalized eigenfunction, the Green's function is given by

$$G(\mathbf{x}, \mathbf{x}') = \sum_n \frac{u_n(\mathbf{x})u_n^*(\mathbf{x}')}{\lambda - \lambda_n} = \frac{2}{L} \sum_n \frac{\sin(n\pi x/L)\sin(n\pi x'/L)}{(n\pi/L)^2 - k^2},$$

where $\lambda = -k^2$ and $\lambda_n = -k_n^2 = -\left(\frac{n\pi}{L}\right)^2$.

6.5.3 A Closed-form Expression of Green's Function for a 1-D Wave Equation

For the 1-D wave equation we can obtain the Green's function in a closed form without using orthogonal expansion (Roach, 1970) as we have already analyzed an acoustic pipe using closed-form solutions. Let us again take the example of string vibration,

$$\frac{d^2u}{dx} + k^2u = -f(x) \qquad 0 \le x \le L,$$

where $f(x)$ denotes distribution of an external force (N) and $u(x)$ shows the displacement of the string vibration (m). If the ends of the string are kept fixed, then this equation must be solved for u subject to boundary conditions such as $u(0) = u(L) = 0$. This type of problem, which requires solving a differential equation subject to specified boundary conditions, is called a boundary value problem.

To solve the boundary value problem posed by the second-order differential equation and associated boundary conditions, we will employ the method of variation of parameters following Roach (1970). We will assume that the solution has the form:

$$u(x) = A(x)\cos kx + B(x)\sin kx.$$

This solution form satisfies the equation

$$-kA'(x)\sin kx + kB'(x)\cos kx = -f(x)$$

subject to

$$0 = A'(x)\cos kx + B'(x)\sin kx,$$

where $A'(x)$ and $B'(x)$ denote differentiation with respect to x, respectively. This can be verified by substituting $u''(x)$ and $u(x)$ into the differential equation

$$\frac{d^2u}{dx} + k^2 u = -f(x).$$

Consequently, we have a set of simultaneous equations for $A'(x)$ and $B'(x)$ as

$$-kA'(x)\sin kx + kB'(x)\cos kx = -f(x)$$

$$0 = A'(x)\cos kx + B'(x)\sin kx.$$

Solving these equations, we get the solutions

$$A'(x) = \frac{f(x)\sin kx}{k}$$

and

$$B'(x) = \frac{-f(x)\cos kx}{k}.$$

Thus we can get the solution formally in the form

$$u(x) = \frac{\cos kx}{k}\int_{c_1}^{x} f(y)\sin ky\,dy - \frac{\sin kx}{k}\int_{c_2}^{x} f(y)\cos ky\,dy,$$

where c_1 and c_2 are constants which are chosen to ensure that the boundary conditions $u(0) = u(L) = 0$ are satisfied.

Introducing one of the boundary conditions, $u(0) = 0$, into the expression of $u(x)$, we obtain

$$u(0) = \frac{1}{k}\int_{c_1}^{0} f(y)\sin ky\,dy = 0.$$

This equation must hold well for any function $f(y)$. Thus, we must choose $c_1 = 0$. The other condition, $u(L) = 0$, requires

$$u(L) = \frac{\cos kL}{k}\int_{0}^{L} f(y)\sin ky\,dy - \frac{\sin kL}{k}\int_{c_2}^{L} f(y)\cos ky\,dy = 0,$$

or equivalently

$$u(L) = \frac{\cos kL}{k}\int_{0}^{L} f(y)\sin ky\,dy - \frac{\sin kL}{k}\int_{0}^{L} f(y)\cos ky\,dy$$

$$+ \frac{\sin kL}{k}\int_{0}^{c_2} f(y)\cos ky\,dy$$

$$= \frac{-1}{k}\int_{0}^{L} f(y)\sin k(L-y)\,dy + \frac{\sin kL}{k}\int_{0}^{c_2} f(y)\cos ky\,dy = 0.$$

Now we can rewrite the solution $u(x)$.

$$u(x) = \frac{\cos kx}{k} \int_{c_1}^{x} f(y) \sin ky \, dy - \frac{\sin kx}{k} \int_{c_2}^{x} f(y) \cos ky \, dy$$

$$= \frac{\cos kx}{k} \int_{0}^{x} f(y) \sin ky \, dy - \frac{\sin kx}{k} \int_{c_2}^{x} f(y) \cos ky \, dy$$

$$= \frac{\cos kx}{k} \int_{0}^{x} f(y) \sin ky \, dy - \frac{\sin kx}{k} \int_{0}^{x} f(y) \cos ky \, dy$$

$$+ \frac{\sin kx}{k} \int_{0}^{c_2} f(y) \cos ky \, dy$$

$$= \frac{-1}{k} \int_{0}^{x} f(y) \sin k(x-y) \, dy + \frac{\sin kx}{k} \int_{0}^{c_2} f(y) \cos ky \, dy$$

$$= \frac{-1}{k} \int_{0}^{x} f(y) \sin k(x-y) \, dy + \frac{\sin kx}{k \sin kL} \int_{0}^{L} f(y) \sin k(L-y) \, dy$$

$$= \frac{-1}{k} \int_{0}^{x} f(y) \sin k(x-y) \, dy + \frac{\sin kx}{k \sin kL} \int_{0}^{x} f(y) \sin k(L-y) \, dy$$

$$+ \frac{\sin kx}{k \sin kL} \int_{x}^{L} f(y) \sin k(L-y) \, dy$$

$$= - \int_{0}^{x} \frac{\sin kL \sin k(x-y) - \sin kx \sin k(L-y)}{k \sin kL} f(y) \, dy$$

$$+ \frac{\sin kx}{k \sin kL} \int_{x}^{L} f(y) \sin k(L-y) \, dy$$

$$= \int_{0}^{x} f(y) \frac{\sin ky \sin k(L-x)}{k \sin kL} \, dy + \int_{x}^{L} f(y) \frac{\sin kx \sin k(L-y)}{k \sin kL} \, dy$$

$$\equiv \int_{0}^{L} f(y) G(x, y) \, dy.$$

Here $G(x, y)$ is the Green's function, which is defined for this case as

$$G(x, y) = \frac{\sin ky \sin k(L-x)}{k \sin kL} \qquad \text{for } 0 \le y \le x$$

and

$$G(x, y) = \frac{\sin kx \sin k(L-y)}{k \sin kL} \qquad \text{for } x \le y \le L.$$

This Green's function exists provided that $\sin kL \ne 0$. Green's function for the string vibration can be expressed in a closed form.

6.5.4 Green's Function and Image Theory

The 1-D Green's function for a string with both ends fixed using an orthogonal expansion form,

$$G(\mathbf{x}, \mathbf{x}') = \sum_n \frac{u_n(\mathbf{x})u_n^*(\mathbf{x}')}{\lambda - \lambda_n} = \frac{2}{L} \sum_n \frac{\sin(n\pi x/L)\sin(n\pi x'/L)}{(n\pi/L)^2 - k^2},$$

can be extended into the 3-D case. A rectangular room which is surrounded by rigid walls provides a fundamental model of the room transfer function and room impulse response (Allen and Berkley, 1979).

Green's function is a solution of

$$\nabla^2 G + k^2 G = -\delta(\mathbf{x} - \mathbf{x}'),$$

where $\mathbf{x}'$ shows the location of a unit point source at (x', y', z'). We take an eigenfunction as

$$\phi_N(\mathbf{x}) = \phi_{l,m,n}(x, y, z) = \cos\left(\frac{l\pi}{L_x}x\right)\cos\left(\frac{m\pi}{L_y}y\right)\cos\left(\frac{n\pi}{L_z}z\right)$$

$$= \cos k_l x \cos k_m y \cos k_n z$$

which satisfies the boundary condition indicating the zero normal velocity at the rigid boundary,

$$\frac{\partial \phi}{\partial n} = 0 \qquad \text{at the boundary,}$$

and L_x, L_y, and L_z are lengths of the sides of the rectangular room, respectively. Then we get

$$G(k, \mathbf{x}, \mathbf{x}') = \sum_N \frac{\Lambda_N}{V} \frac{\phi_N(\mathbf{x})\phi_N(\mathbf{x}')}{k_N^2 - k^2} = \sum_N \frac{\Lambda_N}{V} \frac{\phi_N(\mathbf{x})\phi_N(\mathbf{x}')}{(k_l^2 + k_m^2 + k_n^2) - k^2}$$

$$= \sum_l \sum_m \sum_n \frac{\Lambda_N}{V} \frac{\phi_l(x)\phi_l(x')\phi_m(y)\phi_m(y')\phi_n(z)\phi_n(z')}{(k_l^2 + k_m^2 + k_n^2) - k^2},$$

where

$$\int_V \phi_N^2(\mathbf{x})\, d\mathbf{x} = \frac{V}{\Lambda_N} = \frac{L_x L_y L_z}{\Lambda_N}.$$

$\Lambda_N = \Lambda_{lmn} = 8$ for $l \neq 0, m \neq 0, n \neq 0$ (for oblique waves);
$\Lambda_N = 4$ when one of (l, m, n) is zero for (tangential waves);
and $\Lambda_N = 2$ when two of (l, m, n) are zeros (for axial waves).

This can be confirmed following Section 6.5.1. We use eigenfunction expansions of the forms

$$\delta(\mathbf{x} - \mathbf{x}') = \sum_N d_N(\mathbf{x}')\phi_N(\mathbf{x})$$

and

$$G(k, \mathbf{x}, \mathbf{x}') = \sum_N c_N(\mathbf{x}')\phi_N(\mathbf{x})$$

$$= \sum_N c_N(\mathbf{x}') \cos\left(\frac{l\pi}{L_x}x\right) \cos\left(\frac{m\pi}{L_y}y\right) \cos\left(\frac{n\pi}{L_z}z\right).$$

The expansion coefficient d_N is given by

$$d_N(\mathbf{x}') = \frac{\Lambda_N}{V} \int_V \phi_N(\mathbf{x})\delta(\mathbf{x} - \mathbf{x}')\,d\mathbf{X} = \frac{\Lambda_N}{V}\phi_N(\mathbf{x}').$$

Substituting

$$G(k, \mathbf{x}, \mathbf{x}') = \sum_N c_N(\mathbf{x}')\phi_N(\mathbf{x})$$

into the equation

$$\nabla^2 G + k^2 G = \frac{\partial^2 G}{\partial x^2} + \frac{\partial^2 G}{\partial y^2} + \frac{\partial^2 G}{\partial z^2} + k^2 G = -\delta(\mathbf{x} - \mathbf{x}'),$$

we have

$$-(k_l^2 + k_m^2 + k_n^2)c_N(\mathbf{x}') + k^2 c_N(\mathbf{x}') = -\frac{\Lambda_N}{V}\phi_N(\mathbf{x}').$$

Thus, the coefficient is obtained as

$$c_N(\mathbf{x}') = \frac{\Lambda_N}{V} \frac{\phi_N(\mathbf{x}')}{(k_l^2 + k_m^2 + k_n^2) - k^2} = \frac{\Lambda_N}{V} \frac{\phi_N(\mathbf{x}')}{k_N^2 - k^2}.$$

The Green's function above can be interpreted as the frequency characteristics of the sound field which are produced by an infinite number of image sources in a rectangular room as shown in Fig. 6.5.1. Therefore, we can derive the impulse response of an acoustic system from its Green's function, as we will see in Section 6.5.6. Let us derive a solution in the form of summation of spherical waves which are radiated from the image sources (Allen and Berkley, 1979).

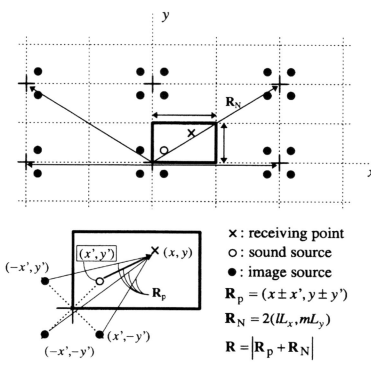

Figure 6.5.1   Two-dimensional display of the image source arrangement for a source in a rectangular room. The actual image space is three-dimensional.

One of the eigenfunctions can be expressed using the exponential expansion as

$$\cos k_l x \cos k_m y \cos k_n z$$

$$= \left(\frac{e^{jk_l x} + e^{-jk_l x}}{2}\right)\left(\frac{e^{jk_m y} + e^{-jk_m y}}{2}\right)\left(\frac{e^{jk_n z} + e^{-jk_n z}}{2}\right).$$

Thus, the product of the eigenfunctions can be written as

$$(\cos k_l x \cos k_m y \cos k_n z)(\cos k_l x' \cos k_m y' \cos k_n z')$$

$$= \left(\frac{e^{jk_l x} + e^{-jk_l x}}{2}\right)\left(\frac{e^{jk_m y} + e^{-jk_m y}}{2}\right)\left(\frac{e^{jk_n z} + e^{-jk_n z}}{2}\right)$$

$$\times \left(\frac{e^{jk_l x'} + e^{-jk_l x'}}{2}\right)\left(\frac{e^{jk_m y'} + e^{-jk_m y'}}{2}\right)\left(\frac{e^{jk_n z'} + e^{-jk_n z'}}{2}\right)$$

$$= \left(\frac{e^{jk_l x} + e^{-jk_l x}}{2}\right)\left(\frac{e^{jk_l x'} + e^{-jk_l x'}}{2}\right)$$

$$\times \left(\frac{e^{jk_m y} + e^{-jk_m y}}{2} \right) \left(\frac{e^{jk_m y'} + e^{-jk_m y'}}{2} \right)$$

$$\times \left(\frac{e^{jk_n z} + e^{-jk_n z}}{2} \right) \left(\frac{e^{jk_n z'} + e^{-jk_n z'}}{2} \right)$$

$$\equiv \left(\frac{e^{jk_l(x\pm x')} + e^{-jk_l(x\mp x')}}{4} \right) \left(\frac{e^{jk_m(y\pm y')} + e^{-jk_m(y\mp y')}}{4} \right)$$

$$\times \left(\frac{e^{jk_n(z\pm z')} + e^{-jk_n(z\mp z')}}{4} \right)$$

$$\equiv \left(\frac{e^{jk_l(x\pm x')\pm jk_m(y\pm y')} + e^{-jk_l(x\mp x')\pm jk_m(y\pm y')}}{16} \right)$$

$$\times \left(\frac{e^{jk_n(z\pm z')} + e^{-jk_n(z\mp z')}}{4} \right)$$

$$\equiv \left(\frac{e^{jk_l(x\pm x')\pm jk_m(y\pm y')\pm jk_n(z\pm z')} + e^{-jk_l(x\mp x')\pm jk_m(y\pm y')\pm jk_n(z\pm z')}}{64} \right)$$

$$\equiv \frac{e^{\pm jk_l(x\pm x')\pm jk_m(y\pm y')\pm jk_n(z\pm z')}}{64} \equiv \sum_{p=1}^{8} \frac{e^{\pm jk_N \cdot R_p}}{64}$$

where we have used $\pm$ notation as an abbreviation for taking a summation required by the inner product of two vectors, and eight vectors are given by the eight permutations over $\pm$ of

$$\mathbf{R}_p = (x \pm x', y \pm y', z \pm z'),$$

whose combinations are assigned from $\mathbf{R}_1$ to $\mathbf{R}_8$, and

$$\mathbf{k}_N \equiv (k_l, k_m, k_n).$$

Then the Green's function can be rewritten as

$$G(k, \mathbf{x}, \mathbf{x}') = \sum_N \frac{\Lambda_N}{V} \frac{\phi_N(\mathbf{x})\phi_N(\mathbf{x}')}{k_N^2 - k^2} = \sum_N \sum_{p=1}^{8} \frac{1}{\Lambda_N V} \frac{e^{-jk_N \cdot R_p}}{k_N^2 - k^2}$$

where $k_N^2 \equiv |\mathbf{k}_N|^2$.

If we introduce the representation of a function using the delta function

$$\int_{-\infty}^{+\infty} f(x)\delta(x - x')\,dx = f(x'),$$

we can use the expression

$$\frac{e^{-j\mathbf{k}_N \cdot \mathbf{R}_p}}{k_N^2 - k^2} = \int\!\!\!\int\!\!\!\int_{-\infty}^{+\infty} \frac{e^{-j\mathbf{u}\cdot\mathbf{R}_p}}{|\mathbf{u}|^2 - k^2}\delta(\mathbf{u} - \mathbf{k}_N)\,du_x\,du_y\,du_z,$$

where

$$|\mathbf{u}|^2 = u_x^2 + u_y^2 + u_z^2$$

and we use the abbreviation

$$\delta(\mathbf{u} - \mathbf{k}_N) = \delta(u_x - k_l)\delta(u_y - k_m)\delta(u_z - k_n).$$

Using this integral representation by the delta function, the Green's function becomes

$$G(k, \mathbf{x}, \mathbf{x}') = \frac{1}{V}\sum_{p=1}^{8}\int\!\!\!\int\!\!\!\int_{-\infty}^{+\infty}\frac{e^{-j\mathbf{u}\cdot\mathbf{R}_p}}{|\mathbf{u}|^2 - k^2}\sum_{N=-\infty}^{+\infty}\frac{1}{\Lambda_N}\delta(\mathbf{u} - \mathbf{k}_N)\,du_x\,du_y\,du_z.$$

Now we recall that a periodic function $f(t)$ with period T can be expressed by a Fourier series, as discussed in Chapter 2. The pulse train

$$\sum_{l=-\infty}^{+\infty}\delta(u_x - k_l) = \sum_{l=-\infty}^{+\infty}\delta\left(u_x - \frac{l\pi}{L_x}\right)$$

is a periodic function with period π/L_x, where $k_l = \pi/L_x$. Thus, we can express the function using the Fourier series as

$$\sum_{l=-\infty}^{+\infty}\delta\left(u_x - \frac{l\pi}{L_x}\right) \equiv \sum_{l=-\infty}^{+\infty}c_l e^{-jl[2\pi/(\pi/L_x)]u_x} = \sum_{l=-\infty}^{+\infty}c_l e^{-j2lL_x u_x}$$

$$= \sum_{l=-\infty}^{+\infty}\frac{L_x}{\pi}e^{-j2lL_x u_x},$$

where

$$c_l = \frac{L_x}{\pi}\int_{-\pi/2L_x}^{\pi/2L_x}\delta\left(u_x - \frac{\pi}{L_x}\right)e^{jl2L_x u_x}\,du_x = \frac{L_x}{\pi}e^{jl2L_x(\pi/L_x)} = \frac{L_x}{\pi}.$$

We have similarly

$$\sum_{m=-\infty}^{+\infty}\delta\left(u_y - \frac{m\pi}{L_y}\right) \equiv \sum_{m=-\infty}^{+\infty}c_m e^{-j2mL_y u_y} = \sum_{m=-\infty}^{+\infty}\frac{L_y}{\pi}e^{-j2mL_y u_y}$$

and

$$\sum_{n=-\infty}^{+\infty} \delta\left(u_z - \frac{n\pi}{L_z}\right) \equiv \sum_{n=-\infty}^{+\infty} c_n e^{-j2nL_z u_z} = \sum_{n=-\infty}^{+\infty} \frac{L_z}{\pi} e^{-j2nL_z u_z}.$$

Thus the Green's function is now expressed as

$$G(k, \mathbf{x}, \mathbf{x}') = \frac{1}{V} \sum_{p=1}^{8} \iiint_{-\infty}^{+\infty} \frac{e^{-j\mathbf{u}\cdot\mathbf{R}_p}}{|\mathbf{u}|^2 - k^2}$$

$$\times \sum_{N=-\infty}^{+\infty} \frac{1}{\Lambda_N} \delta(\mathbf{u} - \mathbf{k}_N)\, du_x\, du_y\, du_z$$

$$= \frac{1}{V} \sum_{p=1}^{8} \iiint_{-\infty}^{+\infty} \frac{e^{-j\mathbf{u}\cdot\mathbf{R}_p}}{|\mathbf{u}|^2 - k^2} \sum_{l=-\infty}^{+\infty} \frac{1}{\Lambda_{lmn}} \delta(u_x - k_l)$$

$$\times \sum_{m=-\infty}^{+\infty} \delta(u_y - k_m) \sum_{n=-\infty}^{+\infty} \delta(u_z - k_n)\, du_x\, du_y\, du_z$$

$$= \frac{1}{V} \sum_{p=1}^{8} \iiint_{-\infty}^{+\infty} \frac{e^{-j\mathbf{u}\cdot\mathbf{R}_p}}{|\mathbf{u}|^2 - k^2} \sum_{l=-\infty}^{+\infty} \frac{1}{\Lambda_{lmn}} \frac{L_x}{\pi} e^{-j2lL_x u_x}$$

$$\times \sum_{m=-\infty}^{+\infty} \frac{L_y}{\pi} e^{-j2mL_y u_y} \sum_{n=-\infty}^{+\infty} e^{-j2nL_z u_z}\, du_x\, du_y\, du_z$$

$$= \sum_{p=1}^{8} \frac{1}{\pi^3} \iiint_{-\infty}^{+\infty} \frac{e^{-j\mathbf{u}\cdot\mathbf{R}_p}}{|\mathbf{u}|^2 - k^2} \sum_{l=-\infty}^{+\infty} \frac{1}{\Lambda_{lmn}}$$

$$\times \sum_{m=-\infty}^{+\infty} \sum_{n=-\infty}^{+\infty} e^{-j2(lL_x u_x + mL_y u_y + nL_z u_z)}\, du_x\, du_y\, du_z$$

$$\equiv \sum_{p=1}^{8} \frac{1}{\pi^3} \iiint_{-\infty}^{+\infty} \frac{e^{-j\mathbf{u}\cdot\mathbf{R}_p}}{|\mathbf{u}|^2 - k^2}$$

$$\times \sum_{N=-\infty}^{+\infty} \frac{1}{\Lambda_N} e^{-j\mathbf{R}_N \cdot \mathbf{u}}\, du_x\, du_y\, du_z$$

$$= \sum_{p=1}^{8} \frac{1}{\pi^3} \iiint_{-\infty}^{+\infty} \sum_{N=-\infty}^{+\infty} \frac{1}{\Lambda_N} \frac{e^{-j\mathbf{u}\cdot(\mathbf{R}_p+\mathbf{R}_N)}}{|\mathbf{u}|^2 - k^2} \, du_x \, du_y \, du_z$$

$$= \sum_{p=1}^{8} \sum_{N=-\infty}^{+\infty} \frac{1}{\Lambda_N \pi^3} \iiint_{-\infty}^{+\infty} \frac{e^{-j\mathbf{u}\cdot(\mathbf{R}_p+\mathbf{R}_N)}}{|\mathbf{u}|^2 - k^2} \, du_x \, du_y \, du_z,$$

where

$$\mathbf{R}_N \equiv 2(lL_x, mL_y, nL_z).$$

Next consider the Fourier transform representation of a spherical wave from a point source in free space. Suppose that there is a point source of unit strength at the origin in free space. We have the wave equation as

$$\nabla^2\phi + k^2\phi = -\delta(\mathbf{x}).$$

We already know that the solution is given by

$$\phi = \frac{e^{-jk|\mathbf{R}|}}{4\pi|\mathbf{R}|},$$

where $|\mathbf{R}|$ is the distance from the source:

$$|\mathbf{R}| = \sqrt{x^2 + y^2 + z^2}.$$

Here we will derive an integral representation of this solution using Fourier transform.

Let us take a solution in the form

$$\phi = \frac{1}{8\pi^3} \iiint_{-\infty}^{+\infty} \Phi e^{-j\mathbf{k}\cdot\mathbf{R}} \, dk_x \, dk_y \, dk_z.$$

Substituting this solution form into the wave equation

$$\nabla^2\phi + k^2\phi = -\delta(\mathbf{x}),$$

we have

$$\frac{-1}{8\pi^3} \iiint_{-\infty}^{+\infty} |\mathbf{k}|^2 \Phi e^{-j\mathbf{k}\cdot\mathbf{R}} \, dk_x \, dk_y \, dk_z$$

$$+ \frac{1}{8\pi^3} k^2 \iiint_{-\infty}^{+\infty} \Phi e^{-j\mathbf{k}\cdot\mathbf{R}} \, dk_x \, dk_y \, dk_z$$

$$= \frac{-1}{8\pi^3} \iiint_{-\infty}^{+\infty} e^{-j\mathbf{k}\cdot\mathbf{R}} \, dk_x \, dk_y \, dk_z,$$

or equivalently,

$$(|\mathbf{k}|^2 - k^2)\Phi = 1,$$

where we have used the relation

$$\delta(\mathbf{x}) = \frac{1}{8\pi^3} \int\!\!\!\int\!\!\!\int\limits_{-\infty}^{+\infty} e^{-j\mathbf{k}\cdot\mathbf{R}}\, dk_x\, dk_y\, dk_z.$$

Hence, we have the representation

$$\phi = \frac{e^{-jk|\mathbf{R}|}}{4\pi|\mathbf{R}|} = \frac{1}{8\pi^3} \int\!\!\!\int\!\!\!\int\limits_{-\infty}^{+\infty} \Phi e^{-j\mathbf{k}\cdot\mathbf{R}}\, dk_x\, dk_y\, dk_z$$

$$= \frac{1}{8\pi^3} \int\!\!\!\int\!\!\!\int\limits_{-\infty}^{+\infty} \frac{e^{-j\mathbf{k}\cdot\mathbf{R}}}{|\mathbf{k}|^2 - k^2}\, dk_x\, dk_y\, dk_z.$$

Consequently, the Green's function

$$G(k, \mathbf{x}, \mathbf{x}') = \sum_{p=1}^{8} \sum_{N=-\infty}^{+\infty} \frac{1}{\Lambda_N \pi^3} \int\!\!\!\int\!\!\!\int\limits_{-\infty}^{+\infty} \frac{e^{-j\mathbf{u}\cdot(\mathbf{R}_p+\mathbf{R}_N)}}{|\mathbf{u}|^2 - k^2}\, du_x\, du_y\, du_z$$

is rewritten as

$$G(k, \mathbf{x}, \mathbf{x}') = \sum_{p=1}^{8} \sum_{N=-\infty}^{+\infty} \frac{8}{\Lambda_N} \frac{e^{-jk|\mathbf{R}_p+\mathbf{R}_N|}}{4\pi|\mathbf{R}_p + \mathbf{R}_N|},$$

where $|\mathbf{R}_p + \mathbf{R}_N|$ denotes the distance between an image (or the original) source and the receiving point as shown in Fig. 6.5.1. Thus, we can see that the Green's function above represents the sound field in a rectangular room by superposition of spherical waves radiated from all of the image sources, as if all the virtual sources were distributed in a free field. $8/\Lambda_N$ denotes the normalized factor for the strength of an image source so that any of the oblique, tangential, and axial modes have unit energy.

Similarly, we can derive Green's function representation using image theory for the 1-D case. Let us take a Green's function for a 1-D acoustic system such as a closed acoustic pipe where an eigenfunction satisfies the boundary condition $\partial\phi/\partial x = 0$ at the closed ends. The Green's function can be written as

$$G(k, x, x')$$

$$= \sum_{l=0}^{+\infty} \frac{u_l(x)u_1^*(x')}{\lambda - \lambda_l} = \frac{2}{L_x} \sum_{l=0}^{+\infty} \frac{\cos(l\pi x/L_x)\cos(l\pi x'/L_x)}{(l\pi/L_x)^2 - k^2}$$

$$= \frac{1}{2L_x} \sum_{l=-\infty}^{+\infty} \frac{e^{-jk_l(x+x')} + e^{-jk_l(x-x')}}{k_l^2 - k^2}$$

$$= \frac{1}{2L_x} \sum_{l=-\infty}^{+\infty} \int_{-\infty}^{+\infty} \frac{e^{-ju_x(x+x')} + e^{-ju_x(x-x')}}{u_x^2 - k^2} \delta(u_x - k_l)\, du_x$$

$$= \frac{1}{2\pi} \int_{-\infty}^{+\infty} \sum_{l=-\infty}^{+\infty} \frac{e^{-ju_x(x+x')} + e^{-ju_x(x-x')}}{u_x^2 - k^2} e^{-j2lL_x u_x}\, du_x$$

$$= \sum_{l=-\infty}^{+\infty} \frac{1}{2\pi} \int_{-\infty}^{+\infty} \frac{e^{-j\{u_x(x+x')+2lL_x u_x\}} + e^{-j\{u_x(x-x')+2lL_x u_x\}}}{u_x^2 - k^2}\, du_x$$

$$= \sum_{l=-\infty}^{+\infty} \frac{1}{2\pi} \int_{-\infty}^{+\infty} \frac{e^{-ju_x\{(x+x')+2lL_x\}} + e^{-ju_x\{(x-x')+2lL_x\}}}{u_x^2 - k^2}\, du_x.$$

Now consider a plane wave solution generated by a source located at the origin which has a unit volume velocity per unit area. A wave solution can be written as

$$\phi = Ae^{-jk|x|}.$$

The coefficient A is determined by the unit volume velocity condition of the source. We have the next relation at the source point:

$$-\left.\frac{\partial \phi}{\partial x}\right|_{x=+0} + \left.\frac{\partial \phi}{\partial x}\right|_{x=-0} = 1.$$

Thus the coefficient A is given by

$$A = \frac{1}{2jk}$$

and the plane wave solution is written as

$$\phi = \frac{1}{2jk} e^{-jk|x|},$$

which satisfies the wave equation

$$\nabla^2 \phi + k^2 \phi = -\delta(x).$$

Therefore, we obtain an integral representation for a plane wave as

$$\frac{1}{2jk} e^{-jk|x|} = \frac{1}{2\pi} \int_{-\infty}^{+\infty} \frac{e^{-jk_x x}}{k_x^2 - k^2}\, dk_x.$$

Consequently, the Green's function becomes

$$G(k, x, x') = \sum_{l=-\infty}^{+\infty} \frac{1}{2\pi} \int_{-\infty}^{+\infty} \frac{e^{-ju_x\{(x+x')+2lL_x\}} + e^{-ju_x\{(x-x')+2lL_x\}}}{u_x^2 - k^2} \, du_x$$

$$= \sum_{l=-\infty}^{+\infty} \frac{e^{-jk|(x+x')+2lL_x|} + e^{-jk|(x-x')+2lL_x|}}{2jk}.$$

6.5.5   Singularity of Green's Function for Free Space

As we described above, the sound field in a rectangular space can be modeled using the image sources, all of which are assumed to be arranged in free space. A wave solution from a unit source in a free field is also called a Green's function in free space which satisfies the source condition. A source is a singularity in a field; however, the singularity characteristics due to the source depends on the dimension of the space.

   We already know a spherical wave solution from a point source in a 3-D free field as

$$\phi = \frac{e^{-jkr}}{4\pi r},$$

which we call Green's function in a 3-D free space. This Green's function has a $1/4\pi r$ type of singularity at the source point. Consider the Green's function for a 2-D free space as a solution of the form

$$\phi \equiv \Phi e^{-jkr}.$$

The unit source condition which has a unit volume velocity can be written as

$$\int_{-\pi}^{+\pi} \left(-\frac{\partial \phi}{\partial r} \right) r \, d\theta = 1,$$

or equivalently

$$\int_{-\pi}^{+\pi} \left(-\frac{\partial \Phi}{\partial r} e^{-jkr} \right) r \, d\theta + \int_{-\pi}^{+\pi} (jk\Phi e^{-jkr}) r \, d\theta$$

$$= \int_{-\pi}^{+\pi} \left(-\frac{\partial \Phi}{\partial r} e^{-jkr} \right) r \, d\theta = 1,$$

where Φ is assumed to be a continuous regular function in our region of interest. If we take the limit as $r \to 0$, we get

$$-\frac{\partial \Phi}{\partial r} 2\pi r = 1,$$

or equivalently

$$-\frac{\partial \Phi}{\partial r} = \frac{1}{2\pi r}.$$

Therefore, we have a solution as

$$\phi = \Phi e^{-jkr} = \frac{-e^{-jkr}}{2\pi} \ln(r) = \frac{e^{-jkr}}{2\pi} \ln\left(\frac{1}{r}\right).$$

The Green's function for a 2-D free space has a $\ln(1/r)$ type of singularity at the unit source location.

For 1-D free space, we have the Green's function

$$\phi = \frac{1}{2jk} e^{-jk|x|}$$

as already shown in Section 6.5.4. This function has no singularity due to the source location, but its derivative has discontinuity at the source point as

$$-\frac{\partial \phi}{\partial x}\bigg|_{x=+0} + \frac{\partial \phi}{\partial x}\bigg|_{x=-0} = 1.$$

6.5.6   Green's Function and Impulse Response

The impulse response of an acoustic system can be derived by inverse Fourier transformation of the Green's function. Let us verify this point by taking the example of the Green's function for a 3-D rectangular room which was derived in Section 6.5.4.

According to Section 6.5.4, the Green's function can be written based on the image method as

$$G(k, \mathbf{x}, \mathbf{x}') = \sum_{N=-\infty}^{+\infty} \frac{\Lambda_N}{V} \frac{\phi_N(\mathbf{x})\phi_N(\mathbf{x}')}{k_N^2 - k^2} = \sum_{p=1}^{8} \sum_{N=-\infty}^{+\infty} \frac{\Lambda_N}{V} \frac{e^{-jk|\mathbf{R}_p + \mathbf{R}_N|}}{4\pi|\mathbf{R}_p + \mathbf{R}_N|}$$

in a 3-D rectangular room with rigid walls. Taking the inverse Fourier transform, we get

$$h(t, \mathbf{x}, \mathbf{x}') = \frac{1}{2\pi} \int_{-\infty}^{+\infty} G(k, \mathbf{x}, \mathbf{x}') e^{j\omega t} \, d\omega$$

$$= \sum_{p=1}^{8} \sum_{N=-\infty}^{+\infty} \frac{8}{\Lambda_N} \frac{1}{2\pi} \int_{-\infty}^{+\infty} \frac{e^{-j(\omega/c)|\mathbf{R}_p + \mathbf{R}_N|}}{4\pi|\mathbf{R}_p + \mathbf{R}_N|} e^{j\omega t} \, d\omega$$

$$= \sum_{p=1}^{8} \sum_{N=-\infty}^{+\infty} \frac{8}{\Lambda_N} \frac{1}{4\pi |\mathbf{R}|_p + \mathbf{R}_N} \frac{1}{2\pi} \int_{-\infty}^{+\infty} e^{j\omega[t-(|\mathbf{R}_p+\mathbf{R}_N|/c)]} \, d\omega$$

$$= \sum_{p=1}^{8} \sum_{N=-\infty}^{+\infty} \frac{8}{\Lambda_N} \frac{\delta\left[t - \dfrac{|\mathbf{R}_p + \mathbf{R}_N|}{c}\right]}{4\pi |\mathbf{R}_p + \mathbf{R}_N|},$$

where we have used the relation

$$\frac{1}{2\pi} \int_{-\infty}^{+\infty} e^{j\omega(t-\tau)} \, d\omega = \delta(t - \tau).$$

Thus, the inverse Fourier transform of the Green's function is the superposition of impulsive spherical waves radiated from all of the image.

The impulse response above shows that the sound field is composed of the direct and reflected waves. We frequently use this decomposition of a sound field into direct and reflected sounds in order to estimate the transfer functions in reverberation spaces. The direct wave is a spherical wave. We can suppose that the sound field must be dominated by the direct wave if we set our receiving position very close to the source position. This will be confirmed analytically in Section 6.5.7.

The impulse response could be derived by taking the inverse Fourier transform of the Green's function. Green's function shows the frequency characteristics function. Therefore, we can use Green's functions whose frequency variables are extended into the complex frequency plane as the transfer function. Thus, the location of the poles of the Green's function is constrained above the real frequency axis.

We will derive the impulse response of an acoustic system in a general form using eigenfunctions. Let us express the impulse response of an acoustic system as

$$h(t, \mathbf{x}, \mathbf{x}') \equiv \frac{1}{2\pi} \int_{-\infty}^{+\infty} G(k, \mathbf{x}, \mathbf{x}') e^{j\omega t} \, d\omega$$

$$= \frac{\Lambda_N}{V} \frac{1}{2\pi} \int_{-\infty}^{+\infty} \sum_N \frac{\phi_N(\mathbf{x})\phi_N(\mathbf{x}')}{k_N^2 - k^2} e^{j\omega t} \, d\omega$$

$$= \frac{\Lambda_N}{V} \frac{1}{2\pi} \sum_N \int_{-\infty}^{+\infty} \frac{\phi_N(\mathbf{x})\phi_N(\mathbf{x}')}{k_N^2 - k^2} e^{j\omega t} \, d\omega$$

$$= \frac{\Lambda_N}{V} \frac{1}{2\pi} \sum_N \phi_N(\mathbf{x})\phi_N(\mathbf{x}') \int_{-\infty}^{+\infty} \frac{e^{j\omega t}}{k_N^2 - k^2} \, d\omega.$$

We will evaluate the integral in the above equation. Here we assume the appropriate system damping δ so that the integral converges. The integral can be written as

$$\int_{-\infty}^{+\infty} \frac{e^{j\omega t}}{k_N^2 - k^2}\, d\omega = c^2 \int_{-\infty}^{+\infty} \frac{e^{j\omega t}}{\omega_N^2 - \omega^2}\, d\omega$$

$$= c^2 \int_{-\infty}^{+\infty} \frac{e^{j\omega t}}{(\omega_N + \omega)(\omega_N - \omega)}\, d\omega$$

$$= -c^2 \int_{-\infty}^{+\infty} \frac{e^{j\omega t}}{(\omega + \omega_N)(\omega - \omega_N)}\, d\omega$$

$$\equiv c^2 \int_{-\infty}^{+\infty} \frac{e^{j\omega t}}{\{\omega - (\omega_{N_0} + j\delta_N)\}\{\omega - (-\omega_{N_0} + j\delta_N)\}}\, d\omega.$$

The singularities of this integrand are located above the real frequency axis as shown in Fig. 6.5.2, since we assume the system damping $\delta_N > 0$.

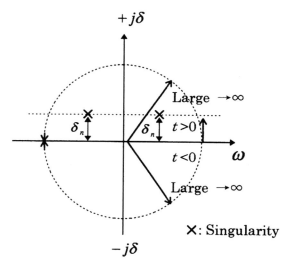

Figure 6.5.2   Singularities of the Green's function and contour for the inverse Fourier transform.

The impulse response is the system response to explosion of an impulsive source. Thus, the acoustic system response must be observed after the explosion. Let us take the time origin at the source explosion event so that the impulse response must be zero in the time region $t < 0$. This is termed *causality* of a linear system. Thus we can take the contour in the lower half-plane as shown in Fig. 6.5.2 for the time region $t < 0$.

The contour integral must be zero, since there are no singularities below the real frequency axis. The integral along the long circular path in Fig. 6.5.2 also converges to zero as the radius of the path becomes large because of the exponential function $e^{+\delta t}$ with negative time ($\delta > 0$). There we can get the integral for the negative time region,

$$c^2 \int_{-\infty}^{+\infty} \frac{e^{j\omega t}}{\{\omega - (\omega_{N_0} + j\delta_N)\}\{\omega - (-\omega_{N_0} + j\delta_N)\}} \, d\omega = 0.$$

For the positive time region, we can take the contour in the upper half-plane as shown in Fig. 6.5.2. We have two singularities in the half plane, but, similarly to the case for the negative time region, the integral along the long circular path converges to zero. Thus we have

$$c^2 \int_{-\infty}^{+\infty} \frac{e^{j\omega t}}{\{\omega - (\omega_{N_0} + j\delta_N)\}\{\omega - (-\omega_{N_0} + j\delta_N)\}} \, d\omega$$

$$= 2\pi j [\mathrm{Res}|_{\omega=(\omega_{N_0}+j\delta_N)} + \mathrm{Res}|_{\omega=(-\omega_{N_0}+j\delta_N)}]$$

$$= 2\pi j c^2 \left[\frac{e^{j(\omega_{N_0}+j\delta_N)t}}{2\omega_{N_0}} - \frac{e^{j(-\omega_{N_0}+j\delta_N)t}}{2\omega_{N_0}} \right]$$

$$= 2\pi j c^2 \left[\frac{e^{j\omega_{N_0}t}}{2\omega_{N_0}} - \frac{e^{-j\omega_{N_0}t}}{2\omega_{N_0}} \right] e^{-\delta_N t}$$

$$= -2\pi c^2 \frac{\sin(\omega_{N_0}t)}{\omega_{N_0}} e^{-\delta_N t} \simeq -2\pi c^2 \frac{\sin(\omega_{N_0}t)}{\omega_{N_0}} \qquad (\delta_N \to 0).$$

Consequently, the impulse response becomes

$$h(t, \mathbf{x}, \mathbf{x}') = \frac{1}{2\pi} \sum_N \frac{\Lambda_N}{V} \phi_N(\mathbf{x})\phi_N(\mathbf{x}') \int_{-\infty}^{+\infty} \frac{e^{j\omega t}}{k_N^2 - k^2} \, d\omega$$

$$= \frac{1}{2\pi} \sum_N \frac{\Lambda_N}{V} \phi_N(\mathbf{x})\phi_N(\mathbf{x}')c^2$$

$$\times \int_{-\infty}^{+\infty} \frac{e^{j\omega t}}{\{\omega - (\omega_{N_0} + j\delta_N)\}\{\omega - (-\omega_{N_0} + j\delta_N)\}} \, d\omega$$

$$= -c^2 \sum_N \frac{\Lambda_N}{V} \phi_N(\mathbf{x})\phi_N(\mathbf{x}') \frac{\sin(\omega_{N_0}t)}{\omega_{N_0}} e^{-\delta_N t}$$

$$\simeq c^2 \sum_N \frac{\Lambda_N}{V} \phi_N(\mathbf{x})\phi_N(\mathbf{x}') \frac{\sin(\omega_{N_0}t)}{\omega_{N_0}} \qquad (t > 0).$$

We can see that the impulse response is composed of free oscillations whose frequencies are eigenfrequencies of the acoustic system of interest. The system damping which we assume so that we can evaluate the impulse response can be estimated by reverberation time of an acoustic system, as we will see in the next chapter. If we introduce appropriate system damping, the impulse response decays exponentially such that

$$h(t, \mathbf{x}, \mathbf{x}') = -c^2 \sum_N \frac{\Lambda_N}{V} \phi_N(\mathbf{x})\phi_N(\mathbf{x}') \frac{\sin(\omega_{N_0}t)}{\omega_{N_0}} e^{-\delta_N t}.$$

If the singularities are located below the real frequency axis as shown in Fig. 6.5.3, then the impulse response becomes noncausal. This is because the integration along the contour in Fig. 6.5.3 has a finite value for the negative time region. Thus, the singularities (or poles) of the transfer function for a causal system must be located in the upper half-plane.

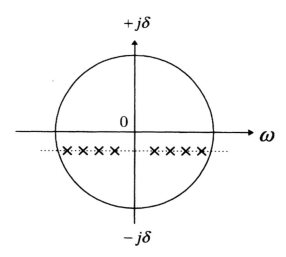

Figure 6.5.3   Singularities of the non-causal impulse response and contour for the inverse Fourier transform.

6.5.7   Direct Field Composed of a Spherical Wave from a Source (Morse and Bolt, 1944)

The Green's function in a rectangular room is different from that in free space, as described in a previous section. However, if we set a receiving position very close to the source point, the sound field around the source can be expected to be almost like a free field. In other words, a spherical wave solution will be seen, since the source effect is more significant on the receiving points than are the boundary effects. We call the areas where we can observe a spherical wave propagation a *direct field*.

Let us derive the direct field using the Green's function. As we have already described in the previous section, the Green's function can be rewritten as

$$G(k, \mathbf{x}, \mathbf{x}') = \sum_N \frac{\Lambda_N}{V} \frac{\phi_N(\mathbf{x})\phi_N(\mathbf{x}')}{k_N^2 - k^2} = \sum_N \frac{\Lambda_N}{V} \frac{\phi_N(\mathbf{x})\phi_N(\mathbf{x}')}{(k_l^2 + k_m^2 + k_n^2) - k^2}$$

$$= \sum_{N=-\infty}^{+\infty} \sum_{p=1}^{8} \frac{1}{\Lambda_N V} \frac{e^{-jk_N \cdot \mathbf{R}_p}}{k_N^2 - k^2},$$

where

$$\phi_N(\mathbf{x}) = \phi_{l,m,n}(x, y, z) = \cos\left(\frac{l\pi}{L_x}x\right) \cos\left(\frac{m\pi}{L_y}y\right) \cos\left(\frac{n\pi}{L_z}z\right)$$

$$= \cos k_l x \cos k_m y \cos k_n z$$

$$\int_V \phi_N^2(\mathbf{x}) \, d\mathbf{x} = \frac{V}{\Lambda_N} = \frac{L_x L_y L_z}{\Lambda_N},$$

$$(\cos k_l x \cos k_m y \cos k_n z)(\cos k_l x' \cos k_m y' \cos k_n z')$$

$$\equiv \frac{e^{\pm jk_l(x \pm x') \pm jk_m(y \pm y') \pm jk_n(z \pm z')}}{64} \equiv \sum_{p=1}^{8} \frac{e^{\pm jk_N \cdot \mathbf{R}_p}}{64},$$

and $\mathbf{R}_p$ represents eight vectors given by the eight permutations over $\pm$ of $\mathbf{R}_p = (x \pm x', y \pm y', z \pm z')$, $\mathbf{k}_N = (k_l, k_m, k_n)$, and $k_N^2 = |\mathbf{k}_N|^2$.

Now we approximate the Green's function given by the infinite series in terms of G_N,

$$G(k, \mathbf{x}, \mathbf{x}') = \sum_{N=-\infty}^{+\infty} \sum_{p=1}^{8} \frac{1}{\Lambda_N V} \frac{e^{-jk_N \cdot \mathbf{R}_p}}{k_N^2 - k^2} \equiv \sum_{N=-\infty}^{+\infty} G_N,$$

using an integral representation. The eigenvalues are located at the lattice points in wavenumber space as shown in Fig. 6.5.4. The Green's function above is summation of all G_N whose eigenvalues are arranged on the lattice points. Thus, we will replace the series summation with respect to the lattice points by taking an integration in the lattice wavenumber space.

For this purpose we impose the conditions that $k \gg (k_{N+1} - k_N)$ and the phase change of $e^{-jk_N \cdot \mathbf{R}_p}$ is within $\pi/2$ from G_N to G_{N+1}. This phase condition requires a slowly changing function for G_N so that we can estimate its integration using samples of G_N. The series of G_N has dominant terms which contribute significantly to the series summation. Suppose that the terms that have their eigenvalues within $k - \alpha \leq k \leq k + \alpha$, which is a thin shell as shown in Fig. 6.5.4, contribute significantly to the series.

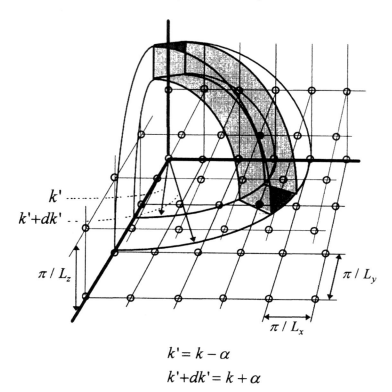

$$k' = k - \alpha$$
$$k' + dk' = k + \alpha$$

Figure 6.5.4 Eigenvalues on the lattice points in 3-D wavenumber space and the thin shell which is assumed to contribute significantly to the series.

The number of eigenvalues within the thin shell can be estimated by

$$\Delta N(k) \equiv n(k)2\alpha = \frac{dN(k)}{dk}2\alpha \cong \frac{V}{\pi^2}k^2\alpha,$$

where

$$n(k) = \frac{dN(k)}{dk} \cong \frac{V}{2\pi^2}k^2$$

as we described in Section 6.4.4. The spherical surface area of the thin shell is given by

$$S_p(k) = \frac{4\pi k^2}{8} = \frac{\pi k^2}{2}.$$

Thus, the mean distance between the eigenvalues can be estimated by

$$\Delta \overline{k_N} \cong \sqrt{\frac{S_p(k)}{\Delta N(k)}} = \sqrt{\frac{\pi k^2/2}{(V/\pi^2)k^2\alpha}} = \sqrt{\frac{\pi^3}{2V\alpha}},$$

where we assume that all the eigenvalues are distributed on the surface, since the shell is thin. Then we get the condition for the distance R from the source by the requirement $\overline{\Delta k_N} R < \pi/2$. That is,

$$\sqrt{\frac{\pi^3}{2V\alpha}} R < \frac{\pi}{2}$$

or equivalently,

$$R < \sqrt{\frac{V\alpha}{2\pi}} \equiv R_c.$$

Suppose that the vector $\mathbf{R} = (x - x', y - y', z - z')$, which indicates the direction and distance between the source and receiving positions, has minimum length of the eight vectors of $\mathbf{R}_p$. Under the condition that this distance satisfies the requirement above,

$$R \equiv |\mathbf{R}| < \sqrt{\frac{V\alpha}{2\pi}} \equiv R_c,$$

we will evaluate the Green's function by integration of

$$G_N = \frac{e^{-j\mathbf{k}_N \cdot \mathbf{R}}}{k_N^2 - k^2}$$

taken over the spherical shell including the negative parts of the lattice space which are located in the other seven octants. Contributions from the other seven $\mathbf{R}_p$ vectors can be included in this integration.

Let us take again a thin shell in the lattice space which includes the lattice points that contribute significantly to the series in Fig. 6.5.4. Suppose that the thin shell has radius k' and its thickness is dk'. Here k' is the variable which is taken as one of the integral variables along the radius of the lattice space. The volume of the thin shell is given by

$$\Delta \equiv 4\pi k'^2 \times dk' = 4\pi k'^2 \, dk'.$$

Since the density of lattice points included in this shell can be estimated by

$$n_\Delta(k') \equiv \frac{8n(k') \, dk'}{\Delta} \cong 8\frac{Vk'^2}{2\pi^2} dk' \bigg/ 4\pi k'^2 \, dk' = \frac{V}{\pi^3},$$

the number of lattice points located in the small section of the thin shell becomes

$$dn_\Delta(k') = \frac{V}{\pi^3} d\Delta = \frac{V}{\pi^3} k'^2 \sin\theta \, d\theta \, d\phi \, dk',$$

where

$$d\Delta \equiv k'^2 \sin\theta \, d\theta \, d\phi \, dk'.$$

The Green's function is estimated as

$$G(k, \mathbf{x}, \mathbf{x}') = \sum_{N=-\infty}^{+\infty} \sum_{p=1}^{8} \frac{1}{\Lambda_N V} \frac{e^{-j\mathbf{k}_N \cdot \mathbf{R}_p}}{k_N^2 - k^2}$$

$$\cong \frac{1}{8V} \int_0^{2\pi} d\phi \int_0^{\pi} \sin\theta \, d\theta \int_0^{\infty} \frac{e^{-jk'|\mathbf{R}|\cos\theta}}{k'^2 - k^2} \frac{V}{\pi^3} k'^2 \, dk'$$

$$= \frac{1}{8\pi^3} \int_0^{2\pi} d\phi \int_0^{\pi} e^{-jk'|\mathbf{R}|\cos\theta} \sin\theta \, d\theta$$

$$\times \int_0^{\infty} \frac{k'^2}{(k'+k)(k'-k)} \, dk',$$

where we take oblique waves only. That is, the approximation

$$n_\Delta(k') \equiv \frac{8n(k')\,dk'}{\Delta} \cong \frac{V}{\pi^3}$$

is introduced only taking account of the number of lattice points which correspond to the oblique waves. We have

$$G(k, \mathbf{x}, \mathbf{x}') \cong \frac{1}{8\pi^3} \int_0^{2\pi} d\phi \int_0^{\pi} e^{-jk'|\mathbf{R}|\cos\theta} \sin\theta \, d\theta$$

$$\times \int_0^{\infty} \frac{k'^2}{(k'-k)(k'+k)} \, dk'.$$

Thus, we can evaluate these integrals as

$$\frac{1}{8\pi^3} \int_0^{2\pi} d\phi = \frac{2\pi}{8\pi^3} = \frac{1}{4\pi^2}$$

and

$$\int_0^{\pi} e^{-jk'|\mathbf{R}|\cos\theta} \sin\theta \, d\theta \int_0^{\infty} \frac{k'^2}{(k'-k)(k'+k)} \, dk'$$

$$= \int_1^{-1} e^{-jk'|\mathbf{R}|x} \sin\theta \frac{dx}{-\sin\theta} \int_0^{\infty} \frac{k'^2}{(k'-k)(k'+k)} \, dk'$$

$$= \int_{-1}^{1} e^{-jk'|\mathbf{R}|x} \, dx \int_0^{\infty} \frac{k'^2}{(k'-k)(k'+k)} \, dk'$$

$$= \int_0^{\infty} \left[\frac{e^{-jk'|\mathbf{R}|} - e^{jk'|\mathbf{R}|}}{-jk'|\mathbf{R}|} \right] \frac{k'^2}{(k'-k)(k'+k)} \, dk'.$$

The final integral representation above can, moreover, be rewritten as

$$\int_0^\infty \left[\frac{e^{-jk'|\mathbf{R}|} - e^{jk'|\mathbf{R}|}}{-jk'|\mathbf{R}|} \right] \frac{k'^2}{(k'-k)(k'+k)} dk'$$

$$\cong \frac{1}{2} \int_{-\infty}^\infty \left[\frac{e^{-jk'|\mathbf{R}|} - e^{jk'|\mathbf{R}|}}{-j|\mathbf{R}|} \right] \frac{dk'}{(k'-k)}$$

if we recall that the integrand has a resonance effect at $k' \cong k$.

Now we will take a limit so that the integration converges. The integral can be divided into two terms. That is,

$$\frac{1}{2} \int_{-\infty}^\infty \left[\frac{e^{-jk'|\mathbf{R}|} - e^{jk'|\mathbf{R}|}}{-j|\mathbf{R}|} \right] \frac{dk'}{(k'-k)}$$

$$\equiv \lim_{\beta_0 \to 0} \frac{1}{2} \int_{-\infty}^\infty \left[\frac{e^{-jk'|\mathbf{R}|} - e^{jk'|\mathbf{R}|}}{-j|\mathbf{R}|} \right] \frac{dk'}{\{k' - (k - j\beta_0)\}}$$

$$= \lim_{\beta_0 \to 0} \frac{-1}{2j|\mathbf{R}|} \int_{-\infty}^\infty \frac{e^{-jk'|\mathbf{R}|} dk'}{\{k' - (k - j\beta_0)\}}$$

$$+ \lim_{\beta_0 \to 0} \frac{1}{2j|\mathbf{R}|} \int_{-\infty}^\infty \frac{e^{jk'|\mathbf{R}|} dk'}{\{k' - (k - j\beta_0)\}}.$$

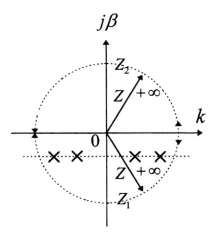

Figure 6.5.5   Example of singularities and integration contours.

The first integral is evaluated in the lower half-plane. Since there is a singularity at $k' = k - j\beta_0$ in the lower half-plane for the first integrand, the first integration on the real wavenumber axis becomes

$$\frac{-1}{2j|\mathbf{R}|} \int_{-\infty}^{\infty} \frac{e^{-jk'|\mathbf{R}|}\,dk'}{\{k'-(k-j\beta_0)\}} = \frac{1}{2j|\mathbf{R}|} 2\pi j[\mathrm{Res}|_{k'=(k-j\beta_0)}]$$

$$= \frac{\pi}{|\mathbf{R}|}\left[e^{-j(k-j\beta_0)|\mathbf{R}|}\right].$$

We take the contour in the upper half-plane in order to evaluate the second integration. We have no singularity in the upper half-plane for the second integrand, and the second integration becomes zero.

Consequently, we have ultimately

$$G(k,\mathbf{x},\mathbf{x}') \cong \frac{1}{8\pi^3} \int_0^{2\pi} d\phi \int_0^{\pi} e^{-jk|\mathbf{R}|\cos\theta}\sin\theta\,d\theta \int_0^{\infty} \frac{k'^2}{(k'-k)(k'+k)}\,dk'$$

$$\cong \frac{1}{4\pi|\mathbf{R}|}\left[e^{-j(k-j\beta_0)|\mathbf{R}|}\right] \cong \frac{1}{4\pi|\mathbf{R}|}\left[e^{-jk|\mathbf{R}|}\right] \qquad (\beta_0 \to 0).$$

This result shows that the Green's function for a finite space can be approximated by a spherical wave in a region which is close to the source position. This region is called a *coherent field* where effects of reflection waves are not significant. We discuss the condition for the coherent field in the next section.

6.5.8   Modal Overlap and Coherent Field (Lyon, 1969)

In the previous section we introduced the condition

$$R < \sqrt{\frac{V\alpha}{2\pi}} \equiv R_c,$$

where R denotes the distance from the source. We will take this as a criterion for the coherent field where the spherical wave from the source is dominant. In the condition above, an unknown quantity α is introduced to represent the number of eigenvalues which make significant contribution to the Green's function expressed by the infinite series

$$G(k,\mathbf{x},\mathbf{x}') = \sum_{N=-\infty}^{+\infty} \sum_{p=1}^{8} \frac{1}{\Lambda_N V} \frac{e^{-j\mathbf{k}_N\cdot\mathbf{R}_p}}{k_N^2 - k^2} \equiv \sum_{N=-\infty}^{+\infty} G_N.$$

Let us evaluate this quantity α. First, we will consider the modal response function, which is given by

$$Y_N \equiv \frac{1}{k_N^2 - k^2} = \frac{1}{(k_N + k)(k_N - k)} \equiv \frac{1}{(k_{N_0} + j\beta_0 + k)(k_{N_0} + j\beta_0 - k)}$$

$$= \frac{1}{k_{N_0}^2 - \beta_0^2 - k^2 + 2j\beta_0 k_{N_0}} \cong \frac{1}{(k_{N_0} + k)(k_{N_0} - k) + 2j\beta_0 k_{N_0}}$$

where $k_N \equiv k_{N_0} + j\beta_0$ and $\beta_0 = c\delta_0$, $\omega_N = ck_N$.
   The squared magnitude becomes

$$|Y_N|^2 \cong \frac{1}{(k_{N_0} + k)^2(k_{N_0} - k)^2 + 4\beta_0^2 k_{N_0}^2}.$$

The maximum is given by

$$|Y_N|_{\max}^2 \cong \frac{1}{4\beta_0^2 k_{N_0}^2}.$$

If we integrate this squared magnitude function, we get

$$\int_0^\infty |Y_N|^2 \, dk \cong \int_0^\infty \frac{1}{(k_{N_0} + k)^2(k_{N_0} + k)^2 + 4\beta_0^2 k_{N_0}^2} \, dk.$$

Similarly to the discussion in Section 6.5.7, if we recall that the integrand has a strong resonance effect at $k_{N_0} \cong k$, this integration can be approximated as

$$\int_0^\infty |Y_N|^2 \, dk \cong \frac{1}{(k_{N_0} + k)^2(k_{N_0} - k)^2 + 4\beta_0^2 k_{N_0}^2} \, dk.$$

$$\cong \int_0^\infty \frac{1}{4k_{N_0}^2(k_{N_0} - k)^2 + 4\beta_0^2 k_{N_0}^2} \, dk.$$

This integration can be evaluated as

$$\int_0^\infty \frac{1}{4k_{N_0}^2(k_{N_0} - k)^2 + 4\beta_0^2 k_{N_0}^2} \, dk = \frac{1}{4k_{N_0}^2 \beta_0^2} \int_0^\infty \frac{1}{(k_{N_0} - k)^2/\beta_0^2 + 1} \, dk$$

$$= \frac{-\beta_0}{4k_{N_0}^2 \beta_0^2} \int_{k_{N_0}/\beta_0}^{-\infty} \frac{1}{\xi^2 + 1} \, d\xi \cong \frac{1}{4k_{N_0}^2 \beta_0} \int_{-\infty}^{\infty} \frac{1}{\xi^2 + 1} \, d\xi$$

where $(k_{N_0} - k)/\beta_0 \equiv \xi$, and

$$\frac{1}{4k_{N_0}^2 \beta_0} \int_{-\infty}^{\infty} \frac{1}{\xi^2 + 1} \, d\xi = \frac{1}{4k_{N_0}^2 \beta_0} [\tan^{-1} \xi]_{-\infty}^{+\infty} = \frac{\pi}{4k_{N_0}^2 \beta_0}$$

$$\cong |Y_N|_{\max}^2 \pi\beta \equiv |Y_N|_{\max}^2 B.$$

Here $B \equiv \pi\beta_0$ is called the modal bandwidth. We can use this modal bandwidth as an estimator of the quantity α, since all the wave modes which have their eigenvalues within this modal bandwidth can be simultaneously excited by a single frequency source, $k \cong k_{N_0}$. The number of eigenvalues included in this modal bandwidth can be estimated by

$$\Delta N(k) = n(k)2\alpha = \frac{dN(k)}{dk}2\alpha \cong \frac{dN(k)}{dk}\pi\beta = \frac{dN(k)}{dk}B.$$

We define the modal overlap M as

$$M = \frac{dN(k)}{dk}B \equiv \frac{dN(k)}{dk}2\alpha \cong \frac{dN(k)}{dk}\pi\beta_0 \cong \frac{V}{2\pi}k^2\beta_0$$

and we have $\alpha \simeq \dfrac{\pi\beta_0}{2}$.

Therefore, we can evaluate the distance R from the source which gives the criterion for the coherent field where the direct sound is dominant as

$$R_c = \sqrt{\frac{V\alpha}{2\pi}} \cong \sqrt{\frac{V\beta_0}{4}}.$$

We can see that the coherent region becomes large as the room volume and/or the system damping β_0 becomes large.

6.5.9 Kirchhoff's Solution for a 3-D Wave Equation (Lamb, 1945)

The transfer functions (TFs) can be represented by Green's functions, but it is not always easy to get the Green's function that satisfies the boundary conditions imposed on the system. Also, the Green's function for free space cannot be applied to a case where the image method is not utilized. In this section we will describe a general method for obtaining a wave solution which satisfies the imposed boundary conditions using a Green's function for free space. This general method is called Kirchhoff's method.

As described in previous sections, we can obtain a solution of the wave equation if we have a Green's function which satisfies the boundary conditions. Following Lamb, we will derive a general solution in a finite region where arbitrary boundary conditions are imposed using a Green's function in free space. This solution form is called Kirchhoff's integral solution for the wave equation. This formula is important as a theoretical basis for sound field control using multiple secondary sources.

Let us again take the wave equation including an external force in 3-D space:

$$\nabla^2\phi + k^2\phi = -\Phi,$$

where Φ is a given function of x, y, z which vanishes outside a finite region Ω as shown in Fig. 6.5.6. We can take a candidate solution as

$$\phi_P(x, y, z) = \frac{1}{4\pi} \iiint_\Omega \Phi(x', y', z') \frac{e^{-jkr}}{r} \, dx' \, dy' \, dz',$$

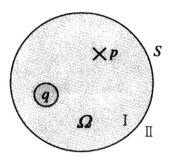

Figure 6.5.6   Source (q) and receiving positions (p) in a finite region.

where r denotes the distance given by

$$r = \sqrt{(x - x')^2 + (y - y')^2 + (z - z')^2}.$$

This can be verified by remembering the solution form using Green's function,

$$\phi(\mathbf{x}) = \int_\Omega G(\mathbf{x}, \mathbf{x}') \Phi(\mathbf{x}') \, d\mathbf{x}'$$

where Green's function is a solution for

$$\nabla^2 G(\mathbf{x}, \mathbf{x}') + k^2 G(\mathbf{x}, \mathbf{x}') = -\delta(\mathbf{x} - \mathbf{x}').$$

A Green's function in 3-D space is given by a spherical wave solution from a point source as

$$G(\mathbf{x}, \mathbf{x}') = \frac{e^{-jkr}}{4\pi r} = \frac{e^{-jk\sqrt{(x-x')^2+(y-y')^2+(z-z')^2}}}{4\pi\sqrt{(x - x')^2 + (y - y')^2 + (z - z')^2}}.$$

If we use this Green's function, then the solution can be written as

$$\phi_P(x, y, z) = \frac{1}{4\pi} \iiint_\Omega \Phi(x', y', z') \frac{e^{-jkr}}{r} \, dx' \, dy' \, dz',$$

However, this is the only solution of the equation

$$\nabla^2 \phi + k^2 \phi = -\Phi$$

which holds at all points of space and vanishes at infinity. In the case of a limited region we may add an arbitrary solution of

$$\nabla^2 \phi + k^2 \phi = 0$$

which satisfies the boundary conditions which are imposed on the boundary S in the finite region Ω. That is, the Green's function

$$G(\mathbf{x}, \mathbf{x}') = \frac{e^{-jkr}}{4\pi r} = \frac{e^{-jk\sqrt{(x-x')^2+(y-y')^2+(z-z')^2}}}{4\pi \sqrt{(x - x')^2 + (z - z')^2 + (z - z')^2}}$$

represents the direct wave from a unit source in 3-D space without the effects of boundaries on the wave propagation.

Now consider a solution for

$$\nabla^2 \phi + k^2 \phi = 0$$

in order to get the solution which satisfies the boundary conditions. Take two functions ϕ_1 and ϕ_2 both of which satisfy the equations, that is,

$$\nabla^2 \phi_1 + k^2 \phi_1 = 0 \qquad \text{and} \qquad \nabla^2 \phi_2 + k^2 \phi_2 = 0.$$

If we recall Green's theorem,

$$\int_S \left(\phi_1 \frac{\partial \phi_2}{\partial n} - \phi_2 \frac{\partial \phi_1}{\partial n} \right) dS = \int_\Omega (\phi_2 \nabla^2 \phi_1 - \phi_1 \nabla^2 \phi_2) \, dv,$$

we get

$$\int_S \left(\phi_1 \frac{\partial \phi_2}{\partial n} - \phi_2 \frac{\partial \phi_1}{\partial n} \right) dS = 0,$$

or equivalently

$$\int_S \phi_1 \frac{\partial \phi_2}{\partial n} \, dS = \int_S \phi_2 \frac{\partial \phi_1}{\partial n} \, dS.$$

We will take a function $\phi_2 = e^{-jkr}/r$ and by setting $\phi_1 = \phi$ we get

$$\int_S \phi \frac{\partial}{\partial n} \left(\frac{e^{-jkr}}{r} \right) dS = \int_S \frac{e^{-jkr}}{r} \frac{\partial \phi}{\partial n} \, dS.$$

Here $\phi_2 = e^{-jkr}/r$ satisfies

$$\nabla^2 \phi_2 + k^2 \phi_2 = 0$$

as already described in Section 6.1.7 except for the case $r = 0$.

Let us take a fixed point p (an observation point p) from which the distance is r used in ϕ_2 representing a spherical wave, e^{-jkr}/r. Suppose that this fixed point is located in the region Ω as shown in Fig. 6.5.7.

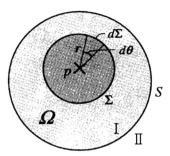

Figure 6.5.7   Small surface encircling the singularity in the region.

Since the function ϕ_2 becomes infinite at the fixed point p, we have to exclude this point from the region to which

$$\int_S \phi \frac{\partial}{\partial n}\left(\frac{e^{-jkr}}{r}\right) dS = \int_S \frac{e^{-jkr}}{r} \frac{\partial \phi}{\partial n} dS \quad \text{and} \quad \nabla^2 \phi_2 + k^2 \phi_2 = 0$$

are required. As Fig. 6.5.7 shows, we divide the region by surrounding the fixed point p with a small spherical surface centered at p which is referred to Σ. Thus, we can rewrite the equation above as

$$\int_\Sigma \phi \frac{\partial}{\partial n}\left(\frac{e^{-jkr}}{r}\right) d\Sigma + \int_S \phi \frac{\partial}{\partial n}\left(\frac{e^{-jkr}}{r}\right) dS$$

$$= \int_\Sigma \frac{e^{-jkr}}{r} \frac{\partial \phi}{\partial n} d\Sigma + \int_S \frac{e^{-jkr}}{r} \frac{\partial \phi}{\partial n} dS.$$

At the surface Σ we have

$$\frac{\partial}{\partial n}\left(\frac{e^{-jkr}}{r}\right) = \frac{\partial}{\partial r}\left(\frac{e^{-jkr}}{r}\right)\frac{\partial r}{\partial n} = \frac{\partial}{\partial r}\left(\frac{e^{-jkr}}{r}\right) = \frac{-jke^{-jkr}r - e^{-jkr}}{r^2}.$$

If we put

$$d\Sigma = r^2 \, d\theta \quad \text{(where } d\theta \text{ denotes the solid angle)}$$

as shown in Fig. 6.5.7 and finally take the limit for $r \to 0$, we have

$$\int_\Sigma \phi \frac{\partial}{\partial n}\left(\frac{e^{-jkr}}{r}\right) d\Sigma$$

$$= \int_0^{4\pi} \phi \frac{-jke^{-jkr}r - e^{-jkr}}{r^2} r^2 \, d\theta$$

$$= \int_0^{4\pi} \phi(-jke^{-jkr}r - e^{-jkr}) \, d\theta \to -4\pi\phi_p \quad (r \to 0),$$

where ϕ_p denotes the value of ϕ at the observation point p, while

$$\int_\Sigma \frac{e^{-jkr}}{r} \frac{\partial\phi}{\partial n} d\Sigma \rightarrow \left(\frac{\partial\phi}{\partial n}\right)\bigg|_p \int_0^{4\pi} \frac{e^{-jkr}}{r} r^2 d\theta \rightarrow 0 \qquad (r \rightarrow 0).$$

Consequently, the relationship

$$\int_\Sigma \phi \frac{\partial}{\partial n} \left(\frac{e^{-jkr}}{r}\right) d\Sigma + \int_S \phi \frac{\partial}{\partial n} \left(\frac{e^{-jkr}}{r}\right) dS$$

$$= \int_\Sigma \frac{e^{-jkr}}{r} \frac{\partial\phi}{\partial n} d\Sigma + \int_S \frac{e^{-jkr}}{r} \frac{\partial\phi}{\partial n} dS$$

becomes, in the limit for $\Sigma \rightarrow 0$,

$$-4\pi\phi_P + \int_S \phi \frac{\partial}{\partial n} \left(\frac{e^{-jkr}}{r}\right) dS = \int_S \frac{e^{-jkr}}{r} \frac{\partial\phi}{\partial n} dS,$$

or equivalently

$$\phi_P = \frac{-1}{4\pi} \int_S \frac{e^{-jkr}}{r} \frac{\partial\phi}{\partial n} dS + \frac{1}{4\pi} \int_S \phi \frac{\partial}{\partial n} \left(\frac{e^{-jkr}}{r}\right) dS.$$

This expression gives the value of ϕ at any point p in a region without sources in terms of the values of ϕ and $\partial\phi/\partial n$ imposed on the boundary by which the region is surrounded. This expression represents a boundary condition using two kinds of source distributions on the boundary. The first term is the velocity potential due to a surface distribution of point sources with a density $-\partial\phi/\partial n$ per unit area, while the second term is the velocity potential of a distribution of double sources with density ϕ whose axes are normal to the surface as described in Section 6.1.7.

Thus, if we combine the two solutions

$$\phi_{p1}(x, y, z) = -\frac{1}{4\pi} \iiint \Phi(x', y', z') \frac{e^{-jkr_1}}{r_1} dx' \, dy' \, dz'$$

and

$$\phi_{P2} = \frac{-1}{4\pi} \int_S \frac{e^{-jkr_2}}{r_2} \frac{\partial\phi}{\partial n} dS + \frac{1}{4\pi} \int_S \phi \frac{\partial}{\partial n} \left(\frac{e^{-jkr_2}}{r_2}\right) dS.$$

then we can obtain the solution as $\phi = \phi_{p1} + \phi_{p2}$ of the wave equation

$$\nabla^2\phi + k^2\phi = -\Phi$$

in a finite region for source and boundary conditions. Here

$$r_1 = \sqrt{(x - x')^2 + (y - y')^2 + (z - z')^2}$$

and similarly r_2 denotes the distance of the point P from the boundaries.

If the (observation) point p is located outside the finite region, we do not have to divide the region by the small surface Σ shown in Fig. 6.5.7, since there are no singular points at which the distance from the point p becomes zero. Thus,

$$\phi_{p2} = \frac{-1}{4\pi} \int_S \frac{e^{-jkr_2}}{r_2} \frac{\partial \phi}{\partial n} dS + \frac{1}{4\pi} \int_S \phi \frac{\partial}{\partial n} \left(\frac{e^{-jkr_2}}{r_2} \right) dS$$

$$= 0$$

since, as we showed in the previous paragraph using Green's theorem,

$$\int_S \phi \frac{\partial}{\partial n} \left(\frac{e^{-jkr}}{r} \right) dS = \int_S \frac{e^{-jkr}}{r} \frac{\partial \phi}{\partial n} dS.$$

We can take one step further and write the expression

$$\phi_{p2} = \frac{-1}{4\pi} \int_S \frac{e^{-jkr_2}}{r_2} \frac{\partial \phi}{\partial n} dS + \frac{1}{4\pi} \int_S \phi \frac{\partial}{\partial n} \left(\frac{e^{-jkr_2}}{r_2} \right) dS$$

at any point of a finite region in terms of point sources only, or double sources only, distributed over the boundary without changing ϕ_{P2} in the region. These variations of boundary distribution expressions are important for sound field control using multiple loudspeakers. Suppose that ϕ_{I} is the velocity potential in the finite region (referred to the first region) and ϕ_{II} denotes the velocity potential outside the finite region (referred to the second region) shown in Fig. 6.5.7. That is, ϕ_{II} satisfies

$$\nabla^2 \phi_{\mathrm{II}} + k^2 \phi_{\mathrm{II}} = 0$$

in the second region and vanishes at infinity in the order of e^{-jkr}/r.

If the point p is located internal to the first region and therefore external to the second, we have

$$\phi_{p2} = \frac{-1}{4\pi} \int_S \frac{e^{-jkr_2}}{r_2} \frac{\partial \phi_{\mathrm{I}}}{\partial n_{\mathrm{I}}} dS + \frac{1}{4\pi} \int_S \phi_{\mathrm{I}} \frac{\partial}{\partial n_{\mathrm{I}}} \left(\frac{e^{-jkr_2}}{r_2} \right) dS$$

and

$$0 = \frac{-1}{4\pi} \int_S \frac{e^{-jkr_2}}{r_2} \frac{\partial \phi_{\mathrm{II}}}{\partial n_{\mathrm{II}}} dS + \frac{1}{4\pi} \int_S \phi_{\mathrm{II}} \frac{\partial}{\partial n_{\mathrm{II}}} \left(\frac{e^{-jkr_2}}{r_2} \right) dS,$$

where $\partial/\partial n_{\mathrm{I}}$ and $\partial/\partial n_{\mathrm{II}}$ denote the normal derivatives to the boundary drawn inward to the first and second regions, respectively, so that

$\partial/\partial n_{II} = -\partial/\partial n_{I}$. Therefore, we can rewrite the second equation as

$$0 = \frac{+1}{4\pi} \int_S \frac{e^{-jkr_2}}{r_2} \frac{\partial \phi_{II}}{\partial n_I} dS - \frac{1}{4\pi} \int_S \phi_{II} \frac{\partial}{\partial n_I} \left(\frac{e^{-jkr_2}}{r_2} \right) dS.$$

By addition of the two equations above, we get

$$\phi_{p2} = \frac{-1}{4\pi} \int_S \frac{e^{-jkr_2}}{r_2} \left(\frac{\partial \phi_I}{\partial n_I} - \frac{\partial \phi_{II}}{\partial n_I} \right) dS$$

$$+ \frac{1}{4\pi} \int_S (\phi_I - \phi_{II}) \frac{\partial}{\partial n_I} \left(\frac{e^{-jkr_2}}{r_2} \right) dS.$$

Consequently, by setting the function ϕ_{II} on the boundary so that

$\phi_{II} = \phi_I$ (continuity of the velocity potential or equivalently
sound pressure at the boundary)

or

$$\frac{\partial \phi_I}{\partial n_I} = \frac{\partial \phi_{II}}{\partial n_I} \qquad \text{(continuity of the particle velocity),}$$

we get the expression

$$\phi_{p2} = \frac{-1}{4\pi} \int_S \frac{e^{-jkr_2}}{r_2} \left(\frac{\partial \phi_I}{\partial n_I} - \frac{\partial \phi_{II}}{\partial n_I} \right) dS \qquad \text{for } \phi_{II} = \phi_I$$

or

$$\phi_{p2} = \frac{1}{4\pi} \int_S (\phi_I - \phi_{II}) \frac{\partial}{\partial n_I} \left(\frac{e^{-jkr_2}}{r_2} \right) dS \qquad \text{for } \frac{\partial \phi_I}{\partial n_I} = \frac{\partial \phi_{II}}{\partial n_I}.$$

The formula obtained by setting that the sound pressure is continuous on the boundary,

$$\phi_{p2} = \frac{-1}{4\pi} \int_S \frac{e^{-jkr_2}}{r_2} \left(\frac{\partial \phi_I}{\partial n_I} - \frac{\partial \phi_{II}}{\partial n_I} \right) dS$$

shows that the sound field in the first region can be expressed using only the distribution of the point sources on the boundary. The density of the point sources is given by

$$-\left(\frac{\partial \phi_I}{\partial n_I} - \frac{\partial \phi_{II}}{\partial n_I} \right).$$

In contrast, the formula derived for the continuity of the particle velocity at the boundary,

$$\phi_{p2} = \frac{1}{4\pi} \int_S (\phi_I - \phi_{II}) \frac{\partial}{\partial n_I} \left(\frac{e^{-jkr_2}}{r_2} \right) dS,$$

expresses the sound field in the first region using only double-source distribution on the boundary. The density of the double sources is given by $\phi_I - \phi_{II}$.

The derivation above can be extended to the form including the time factor. The original formula

$$\phi_{p2} = \frac{-1}{4\pi} \int_S \frac{e^{-jkr_2}}{r_2} \frac{\partial\phi}{\partial n} dS + \frac{1}{4\pi} \int_S \phi \frac{\partial}{\partial n} \left(\frac{e^{-jkr_2}}{r_2} \right) dS$$

can be extended to

$$\phi_{p2} = \frac{-1}{4\pi} \int_S \frac{e^{j\omega(t-r_2/c)}}{r_2} \frac{\partial\phi}{\partial n} dS + \frac{1}{4\pi} \int_S \phi \frac{\partial}{\partial n} \left(\frac{e^{j\omega(t-r_2/c)}}{r_2} \right) dS,$$

where c denotes the sound speed. If we recall that a signal can be represented using the Fourier transform

$$f(t) = \frac{1}{2\pi} \int F(\omega)e^{j\omega t} d\omega,$$

then we can generalize the stationary sinusoidal signal form. This is called Kirchhoff's integral representation of the wave solution.

6.6 Summary

Wave physics is basic science for acoustic systems to which signal processing is applied. This chapter summarized the wave properties of linear acoustic systems that provide an insight into acoustic data structures. Wave equations, sound radiation and propagation, acoustic impedance, the image method, eigenvalues and eigenfunctions, Hermite operators and orthogonal expansions, Green's functions, transfer functions and impulse responses, and Kirchhoff's solution are basic notions in wave physics. Acoustic systems are characterized by Green's functions. Green's functions are the same as transfer functions in linear systems, since the inverse Fourier transform of Green's function is the impulse response of the system. Green's functions have spatial, time, and frequency variables. Green's function for a finite system such as a rectangular room has poles and zeros, and the poles are distributed nonuniformly and rather randomly in the complex frequency domain. It is not always possible to get a simple form of Green's function. Kirchhoff's method provides a wave solution which satisfies boundary conditions using Green's function for free space.

7 Statistical Models for Acoustic Transfer Functions

7.0 Introduction

We described transfer function (TF) models based on the wave equation in the previous chapter. Our acoustic systems such as rooms can be described in a deterministic way following the wave equation. The sound fields in acoustic systems, however, are sensitive to the sound source and receiving positions or to the boundary conditions. Since it seems impractical to characterize the complicated TFs based on the wave theoretic approach, a statistical approach using a few parameters has been taken as a practical way to estimate the TF. In this chapter, we will describe the fundamental properties for the TF models of random sound fields. The random sound fields can be modeled by superposition of plane waves with random magnitude and phase. Such a model simulates superposition of direct sound and reflections whose magnitudes and delay times are random. Thus, this provides a polynomial model of the TF which has random coefficients. The polynomial model of TFs is represented by the distribution of zeros instead of poles. Poles are produced in the limit, when the order of the polynomials becomes infinity. The statistics of the distribution of zeros are investigated as well as the magnitude and phase of the TF.

7.1 Random Sound Field Model

7.1.1 Superposition of Random Reflections

A sound field in a closed space is composed of the direct sound and reflections from the boundaries. Reverberation sound fields with a lot of reflection waves from different directions can be formulated as

$$p(t) = p_0 \sum_{i=1}^{N} \cos(\omega t + \alpha_i) \quad \text{(Pa)},$$

where ω is the angular frequency of the sound (rad s^{-1}), N denotes the number of waves to be superposed, and all the magnitudes of these waves are assumed to be p_0. The equation shows the summation of sinusoidal waves of the same frequency with different phases. The mean square pressure becomes

$$\overline{p^2}(t) = \tfrac{1}{2} p_0^2 (A^2 + B^2) \quad (\text{Pa}^2),$$

where

$$A \equiv \sum_{i=1}^{N} \cos \alpha_i \qquad \text{and} \qquad B \equiv \sum_{i=1}^{N} \sin \alpha_i.$$

Here the overbar denotes taking the time average.

Suppose that the phase α_i is a random variable uniformly distributed from 0 to 2π. Take an ensemble of the sound pressure data which are randomly sampled at many positions in the space. When α_i randomly takes a value between 0 and 2π, the samples of $\cos \alpha_i$ and $\sin \alpha_i$ are distributed between the limits ± 1. But the distributions of both $\sum_{i=1}^{N} \cos \alpha_i$ and $\sum_{i=1}^{N} \sin \alpha_i$ in the ensemble approach the normal distribution by the central limit theorem, as N increases. Figure 7.1.1 shows that the distributions approach the normal distribution.

These results can be extended to the case where all the waves have different magnitudes. The sound pressure can be formulated as

$$p(t) = \sum_{i=1}^{N} p_i \cos(\omega t + \alpha_i) \quad (\text{Pa})$$

$$= \left[\sum_{i=1}^{N} p_i \cos(\alpha_i) \right] \cos(\omega t) - \left[\sum_{i=1}^{N} p_i \sin(\alpha_i) \right] \sin(\omega t)$$

$$\equiv A \cos(\omega t) - B \sin(\omega t),$$

where

$$A \equiv \left[\sum_{i=1}^{N} p_i \cos(\alpha_i) \right] \qquad \text{and} \qquad B \equiv \left[\sum_{i=1}^{N} p_i \sin(\alpha_i) \right].$$

Thus, we have again

$$\overline{p^2}(t) = \tfrac{1}{2} p_0^2 (A^2 + B^2) \quad (\text{Pa}^2)$$

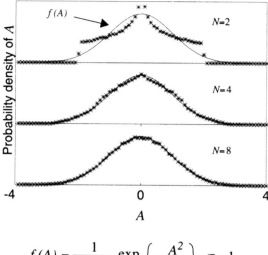

$$f(A) = \frac{1}{\sqrt{2\pi\sigma^2}} \exp\left\{-\frac{A^2}{2\sigma^2}\right\}, \quad \sigma = 1$$

α_i: Uniform distribution $(0 \sim 2\pi)$

Number of samples: 10^5

Figure 7.1.1   Distributions of $\sum_{i=1}^{N} \cos\alpha_i$.

We can expect that the distributions of both $\sum_{i=1}^{N} p_i \cos\alpha_i$ and $\sum_{i=1}^{N} p_i \sin\alpha_i$ in the ensemble approach the normal distribution by the central limit theorem, as N increases.

7.1.2   Rayleigh and Exponential Distributions

Let us take two independent Gaussian variables x and y both of which have 0 mean and the variance σ^2. The probability density functions are written as

$$p(x) = \frac{1}{\sqrt{2\pi\sigma^2}} \exp\left(-\frac{x^2}{2\sigma^2}\right)$$

and

$$p(y) = \frac{1}{\sqrt{2\pi\sigma^2}} \exp\left(-\frac{y^2}{2\sigma^2}\right).$$

The combined variable $z = \sqrt{x^2 + y^2}$ is known to have the probability density function called a Rayleigh distribution.

The probability that the variable z takes a value within z and $z + dz$ is proportional to the area of the ring, as shown in Fig. 7.1.2. Changing the variables as

$$x = z \cos \theta \quad \text{and} \quad y = z \sin \theta,$$

we get the simultaneous probability as

$$p(x, y)\, dx\, dy = p(z \cos \theta, z \sin \theta) z\, dz\, d\theta$$

$$= p(x)p(y)\, dx\, dy = \frac{1}{2\pi\sigma^2} \exp \left\{ -\frac{z^2}{2\sigma^2} \right\} z\, dz\, d\theta.$$

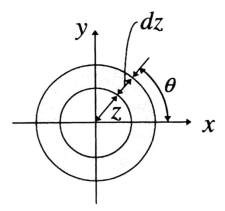

Figure 7.1.2   Variables z and $z + dz$.

Therefore, we have

$$p(z)\, dz = \int_0^{2\pi} p(z \cos \theta, z \sin \theta) z\, dz\, d\theta$$

$$= \frac{1}{2\pi\sigma^2} \int_0^{2\pi} \exp \left\{ -\frac{z^2}{2\sigma^2} \right\} z\, dz\, d\theta = \frac{z\, dz}{\sigma^2} \exp \left\{ -\frac{z^2}{2\sigma^2} \right\}$$

or

$$p(z) = \frac{z}{\sigma^2} \exp \left\{ -\frac{z^2}{2\sigma^2} \right\}.$$

This probability density function is called a Rayleigh distribution.

If we again change the variable, we get the probability density function for the squared quantity, such as mean squared pressure. Let us put

$$z^2 = x^2 + y^2 \equiv w$$

or

$$z = \sqrt{x^2 + y^2} = \sqrt{w}.$$

We have

$$p(w)\,dw = \frac{\sqrt{w}}{\sigma^2} \exp\left\{-\frac{w}{2\sigma^2}\right\} \frac{dw}{2\sqrt{w}} = \frac{1}{2\sigma^2} \exp\left\{-\frac{w}{2\sigma^2}\right\} dw$$

or

$$p(w) = \frac{1}{2\sigma^2} \exp\left\{-\frac{w}{2\sigma^2}\right\}$$

This is called an exponential distribution. The random variable of the squared sum of two uncorrelated Gaussian variables follows an exponential distribution.

7.1.3 Distribution of Magnitude and Square Magnitude

The square magnitude of the sound pressure, that is, the mean squared pressure, is given by

$$\overline{p^2}(t) = \tfrac{1}{2} p_0^2 (A^2 + B^2) \quad (\text{Pa}^2),$$

where

$$A \equiv \sum_{i=1}^{N} \cos \alpha_i \quad \text{and} \quad B \equiv \sum_{i=1}^{N} \sin \alpha_i.$$

The variance of the random variable composed of finite number of uncorrelated random components is given by the summation of the variance for each component. For example, suppose a random variable such as $z = x + y$ where x and y are uncorrelated random variables of zero mean. The variance, which is given by the expectation of the squared variable, is written as

$$E[z^2] = E[(x+y)^2] = E(x^2) + E(y^2) + 2E(xy) = E(x^2) + E(y^2).$$

Here $E[*]$ denotes taking the expectation and $E(xy) = 0$, since x and y are uncorrelated with each other.

Let us take a random variable $x = \cos \alpha$. Suppose that α takes with equal probability any value between 0 and 2π. The expectation of x^2 can be written as

$$E[x^2] = \frac{1}{2\pi} \int_0^{2\pi} \cos^2 \alpha \, d\alpha = \frac{1}{2}.$$

Similarly, we get

$$E[y^2] = \frac{1}{2\pi} \int_0^{2\pi} \sin^2 \alpha \, d\alpha = \frac{1}{2}$$

for a random variable such as $y = \sin \alpha$ Therefore, both $A = \sum_{i=1}^{N} \cos \alpha_i$ and $B = \sum_{i=1}^{N} \sin \alpha_i$ have variance of $N/2$.

The random variable of the squared sum of two uncorrelated Gaussian variables follows an exponential distribution as stated in Section 7.1.2. Figure 7.1.3 demonstrates an example of mean squared sound pressure distributions. We can confirm the exponential distribution.

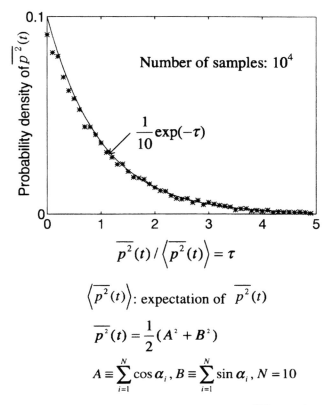

$$\overline{p^2}(t) / \left\langle \overline{p^2}(t) \right\rangle = \tau$$

$$\left\langle \overline{p^2}(t) \right\rangle : \text{expectation of } \overline{p^2}(t)$$

$$\overline{p^2}(t) = \frac{1}{2}(A^2 + B^2)$$

$$A \equiv \sum_{i=1}^{N} \cos \alpha_i, B \equiv \sum_{i=1}^{N} \sin \alpha_i, N = 10$$

Figure 7.1.3   Distribution of mean squared pressure $\overline{p^2}(t) = \frac{1}{2}P_0^2(A^2 + B^2)$, where $A \equiv \sum_{i=1}^{N} \cos \alpha_i$ and $B \equiv \sum_{i=1}^{N} \sin \alpha_i$.

7.2   Random Polynomial Model of the Transfer Function

7.2.1   Combined Waves for Direct Wave and Reflections

Random sound fields which are composed of direct and reflection waves provide a fundamental model of finite impulse response systems. Suppose that we have a direct wave and a reflection with a delay of T (s). The impulse response for such a system can be written as

$$h(t) = \delta(t) + A\delta(t - T)$$

as shown in Fig. 7.2.1. Here $\delta(t)$ is the delta function, and A corresponds the reflection coefficient of a wall which produces the single reflection. If we take a sinusoidal wave of a single frequency with unit amplitude as an input signal, the TF which represents the frequency characteristic of a system is expressed as the phasor (complex magnitude) of the output signal $y(t)$ of the system. The output signal is written as

$$y(t) \equiv H(\omega)e^{j\omega t} = \left(1 + Ae^{-j\omega T}\right)e^{j\omega t},$$

where

$$H(\omega) \equiv \int_{-\infty}^{+\infty} h(t)\exp(-j\omega t)\,dt$$

$$= \int_{-\infty}^{+\infty} (\delta(t) + A\delta(t - T))\exp(-j\omega t)\,dt$$

$$= 1 + Ae^{-j\omega T}.$$

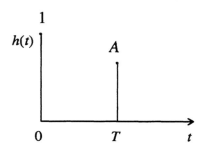

Figure 7.2.1 Impulse response of the direct and a single reflection wave.

Figure 7.2.2 illustrates examples of the magnitude of the TF under various conditions of A. The TF stated above can be characterized by the periodic distribution of zeros on the real frequency axis, when $A = 1$. The squared magnitude (squared sum of the real and imaginary parts) of the TF becomes

$$|H(\omega)|^2 = |1 + Ae^{-j\omega T}|^2$$

$$= (1 + A\cos(\omega T))^2 + A^2\sin^2(\omega T) = (1 + A^2) + 2A\cos(\omega T).$$

The frequency which makes the magnitude zero is called zero for short. When $A = 1$, such zeros ω_0 for the TF stated above must satisfy the equation,

$$|H(\omega_0)|^2 = (1 + 1) + 2\cos(\omega_0 T) = 0.$$

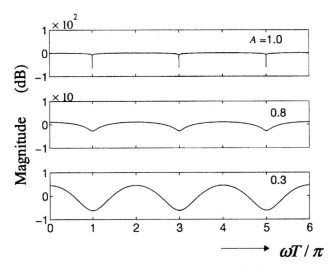

Figure 7.2.2 Magnitudes of the TFs for various reflection wave amplitudes A.

Therefore, we can get

$$\cos(\omega_0 T) = -1.$$

We have the frequency which makes the magnitude zero,

$$\omega_0 T = \frac{2n - 1}{2}\pi \qquad n = 1, 2, 3, \ldots.$$

Such frequencies are located periodically.

7.2.2 Polynomials for the Transfer Functions of Random Sound Fields

The TF for the direct wave and the reflection with delay T,

$$H(\omega) = 1 + Ae^{-j\omega T}$$

can be extended into the complex frequency domain by introducing the new variable z as

$$z \equiv e^{j\omega T} \equiv e^{j(\omega_r + j\delta)T}.$$

Here the frequency is extended into the complex frequency plane as

$$\omega \equiv \omega_r + j\delta.$$

The TF can be rewritten as

$$H(z) = 1 + Az^{-1}.$$

This is just same as taking the z-transform (Chapter 3) of the impulse response

$$h(t) = \delta(t) + A\delta(t - T).$$

If we take the z-transform of the impulse response which is composed of finite number of reflections, the TF for the random sound fields composed of N reflections can be modeled by a polynomial as

$$H(z) = \sum_{n=0}^{N} a_n z^{-n}$$

where $z = e^{j\omega T}$, T denotes the unit delay of the reflection waves, and the coefficient a_n shows the magnitude of reflection sound which has the nth unit delay time. The zeros are complex roots of the TF. The TF given by the polynomial $H(z)$ is characterized by the zeros instead of the poles. An Nth order polynomial has N zeros.

7.2.3   Zeros of Random Polynomials

Figure 7.2.3 shows examples of distributions of zeros for the polynomial $H(z) = \sum_{n=0}^{N} a_n z^{-n}$ in the complex frequency plane $\omega + j\delta$ where $-\pi \leq \omega T < \pi$. The polynomial $H(z)$ is a periodic function of ωT on the unit circle of $z = e^{j\omega T}$. The period is 2π since $z = e^{j\omega T} = e^{j(\omega T + 2\pi)}$.

Suppose that a_n is a real random variable which follows a normal distribution. A set of a_n represents random amplitudes of reflections. Figure 7.2.3a plots all the zeros of 20 random coefficient polynomials of order 10. The zero locations are concentrated in the area close to the real frequency axis. Figure 7.2.3b is the density of zeros calculated from Fig. 7.2.3a. The density of zeros becomes high as δ, which is the distance from the real frequency axis, becomes small. Figure 7.2.3c shows the accumulated number of zeros up to δ from the far end of negative δ. The distribution follows essentially a power law in δ, as shown by the solid line in the figure.

To simulate the decay process of room impulse responses, we apply an exponential windowing function to the random polynomials. In terms of statistics, the variance of a_n decreases exponentially as the order of the coefficients increases. Figure 7.2.4 illustrates samples of the distribution of zeros. The distribution patterns are shifted into the positive δ direction by the decay parameter δ_0 of the exponential windowing function $W \equiv e^{-\delta_0 n T}$. The density of the zeros is concentrated on the line $\delta = \delta_0$.

The density is concentrated on the line which is estimated by the reverberation time. The high density line of the zeros can be estimated by

$$\delta = \delta_0 \cong \frac{6.9}{T_R} \quad (\text{s}^{-1}).$$

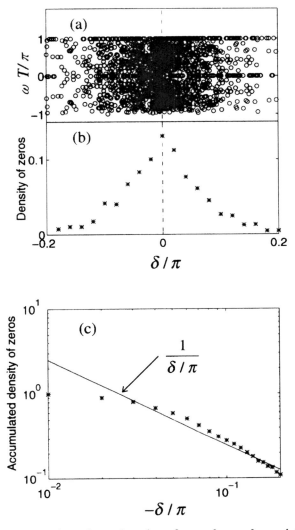

Figure 7.2.3 Distribution of zero locations for random polynomials with coefficients of Gaussian variables. (a) Distribution of zeros for 20 random polynomials of order 10. (b) Density of zeros for the distribution of (a). (c) Accumulated number of zeros in the complex frequency plane up to δ from the far end of negative δ.

The reverberation time T_R is defined as the 60 dB decay time in the sound energy decay process. If we take the energy decay process following $E(t) = e^{-2\delta_0 t}$, then the reverberation time is obtained by solving

$$10 \log(e^{-2\delta_0 T_R}) = -60 \quad \text{(dB)}.$$

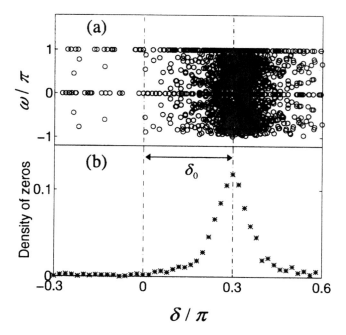

Figure 7.2.4 Distribution of zeros for exponentially weighted random polynomials. (a) Distribution of zeros for 20 random polynomials of order 10. (b) Density of zeros for the distribution of (a).

Consequently, we get

$$T_R = \frac{3}{\delta_0} \log_{10} e \cong \frac{6.9}{\delta_0} \quad \text{(s)}.$$

The impulse response based on the wave equation model is formulated by an exponentially decaying function which can be interpreted as the limit case of the polynomial models when $N \to \infty$. If we again take such a wave equation model for the random TFs, then the occurrence of zeros can be investigated analytically using poles and residues as described later.

7.3 Reverberation Statistics in Rooms

7.3.1 Number of Reflection Waves

Figure 7.3.1 is an example of the arrangement of image sources in a two-dimensional case. A reflection sound can be represented by a spherical wave which is radiated from an image source. Let us consider the number of reflection sounds $N(t)$ arriving at t (s) after the decay process started.

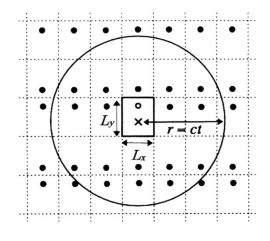

O : source c : sound speed

● : image source   ✕ : observer

Figure 7.3.1   Source location and image sources.

The number of reflection sounds can be estimated by the number of image sources which are located in a sphere of radius ct from the receiving position. Here c is the sound speed. That is,

$$N(t) \cong \frac{4\pi(ct)^3}{3V},$$

where V denotes the room volume (m^3). The density $n(t)$ is given by taking the first derivative of t:

$$n(t) \cong \frac{4\pi c^3 t^2}{V}.$$

The number of reflection waves arriving at the receiving position in a unit time interval increases in proportion to the square of time as the decay process goes on.

7.3.2   Number of Collisions at the Boundaries

The average number of collisions at the boundaries after t (s) since the decay process started can be estimated. Let us take polar coordinates as shown in Fig. 7.3.2. The reflection sound from an image source located at (ct, θ, ϕ) in the figure undergoes the collisions at the boundaries until it arrives at the receiving point. The distance between the receiving point and the image source is ct. The number of collisions at x-walls can be

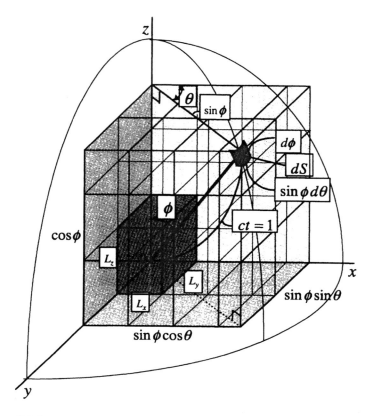

Figure 7.3.2   Polar and rectangular coordinates for image source locations.

estimated as

$$R_x \cong \frac{ct \sin \phi \cos \theta}{L_x}.$$

Similarly, we get

$$R_y \cong \frac{ct \sin \phi \sin \theta}{L_y}$$

for y-wall collisions, and

$$R_z \cong \frac{ct \cos \phi}{L_z}$$

for z-wall collisions, where L_x, L_y, and L_z denote the lengths of the sides, respectively. Thus the total number of collisions becomes

$$R \equiv R_x + R_y + R_z \cong \frac{ct \sin \phi \cos \theta}{L_x} + \frac{ct \sin \phi \sin \theta}{L_y} + \frac{ct \cos \phi}{L_z}.$$

The number of collisions depends on the geometric relation between the image source locations and receiving points. Thus we estimate an expectation for the collision number by randomly sampling receiving positions. As shown in Fig. 7.3.2, the probability density $p(\theta, \phi)$ for the image source distribution is given by the ratio of the infinitesimally small area

$$dS = \sin \phi \, d\phi \, d\theta$$

and the 1/8 surface area of a unit sphere

$$S = \frac{4\pi}{8} = \frac{\pi}{2}.$$

Thus, we get

$$p(\theta, \phi) \cong \frac{dS}{S} = \frac{\sin \phi \, d\phi \, d\theta}{\pi/2}.$$

Consequently, we have the expectation as

$$E(R) \cong \overline{R} = \overline{R_x} + \overline{R_y} + \overline{R_z} = \frac{ct}{2L_x} + \frac{ct}{2L_y} + \frac{ct}{2L_z}$$

$$= \frac{L_xL_y + L_yL_z + L_zL_x}{2L_xL_yL_z}ct = \frac{cSt}{4V} \equiv \frac{ct}{m},$$

where

$$\overline{R} = \frac{2}{\pi} \int_0^{\pi/2} \int_0^{\pi/2} \left(R_x + R_y + R_z\right) \sin \phi \, d\phi \, d\theta,$$

$$\overline{R_x} = \frac{2}{\pi} \int_0^{\pi/2} \int_0^{\pi/2} \frac{ct \sin \phi \cos \theta}{L_x} \sin \phi \, d\phi \, d\theta$$

$$= \frac{2}{\pi} \frac{ct}{L_x} \int_0^{\pi/2} \sin^2 \phi \, d\phi \int_0^{\pi/2} \cos \theta \, d\theta = \frac{ct}{2L_x},$$

$$\overline{R_y} = \frac{ct}{2L_y} \qquad \overline{R_z} = \frac{ct}{2L_z},$$

and S (m^2) denotes the surface areas. Here

$$m \equiv \frac{4V}{S} \quad \text{(m)}$$

is called the mean free path, which is the average distance that the sound travels between two successive collisions.

7.3.3   Energy Decay Process

The intensity of a spherical wave is given by

$$J_0(r) = \frac{P_0}{4\pi r^2} \quad (\text{W m}^{-2})$$

at a distance r (m) from the point source with power output P_0 (W). The intensity of the reflection sounds for image sources in a unit time interval is expressed as

$$J(t) = \frac{P_0(1-\alpha)^{R(t)}}{4\pi(ct)^2} n(t) \cong \frac{P_0(1-\alpha)^{R(t)}c}{V} \quad (\text{W m}^{-2}),$$

where

$$n(t) \cong \frac{4\pi c^3 t^2}{V},$$

and α denotes the average absorption coefficients of the space as

$$\alpha = \frac{\sum_i \alpha_i S_i}{S}.$$

α_i is the absorption coefficient of the ith portion of the surface whose area is S_i (m^2) and the total area is S (m^2). If we substitute the average number of collisions $\overline{R}(t)$,

$$\overline{R}(t) = \frac{cSt}{4V}$$

instead of $R(t)$, we get

$$\overline{J}(t) \equiv \frac{P_0(1-\alpha)^{\overline{R}(t)}c}{V} = \frac{P_0(1-\alpha)^{cSt/4V}c}{V} \quad (\text{W m}^{-2})$$

in the unit time interval.
Rewriting,

$$(1-\alpha)^{cSt/4V} = \exp\left(\frac{cS}{4V}[\ln(1-\alpha)]t\right)$$

we obtain

$$\overline{J}(t) = \frac{P_0(1-\alpha)^{cSt/4V}c}{V}$$

$$= \frac{P_0 c}{V} \exp\left(\frac{cS}{4V}[\ln(1-\alpha)]t\right) \quad (\text{W m}^{-2}).$$

The reverberation time T_R is obtained from the equation

$$10 \log \frac{\overline{J}(T_R)}{\overline{J}(0)} = -60 \quad (\text{dB})$$

as

$$T_R = \frac{0.161V}{-\ln (1 - \alpha)S} \quad (\text{s}).$$

7.4 Poles and Zeros of Room Transfer Function Models

7.4.1 Poles and Zeros

In Section 7.2 we described the zeros of the random polynomials that model the TFs for sound fields which are composed of a finite number of random reflections. In this section we investigate the distribution statistics of poles and zeros of Green's functions from a wave model for random sound fields. If we extend the real frequency ω into the complex variable $\omega + j\delta$, the TF can be identified as a complex function by the locations of poles and zeros in the complex frequency domain as shown in Fig. 7.4.1.

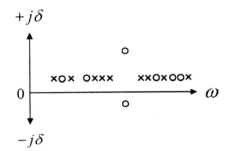

Figure 7.4.1 Complex frequency plane.

When we describe the TF as a rational function,

$$G(k, \mathbf{x}, \mathbf{x}') = \sum_N \frac{u_N(\mathbf{x})u_N^*(\mathbf{x}')}{k_N^2 - k^2} \equiv \sum_N \frac{a_N(\mathbf{x}, \mathbf{x}')}{k_N^2 - k^2}$$

$$\equiv \frac{N(\omega, \mathbf{x}, \mathbf{x}')}{D(\omega, \mathbf{x}, \mathbf{x}')} \equiv H(\omega, \mathbf{x}, \mathbf{x}').$$

Here the poles are singularities (the complex roots of the denominator) in the complex frequency plane and the zeros are complex roots of the TF. If we assume $\exp(j\omega t)$ time dependency for input and output sinusoidal signals, the poles are located in the upper half-plane of the complex frequency domain because of the causality of acoustic systems.

If a pole is located at

$$\omega_p = \omega_{pr} - j\delta(\delta > 0)$$

in the lower half-plane, then the free oscillation which composes the impulse response obtained by the inverse Fourier transform of

$$H_p(\omega) = \frac{1}{\omega - \omega_p}$$

becomes

$$h(t) = \frac{1}{2\pi} \int_{-\infty}^{+\infty} \frac{1}{\omega - \omega_p} e^{j\omega t} \, d\omega$$

$$= \frac{1}{2\pi}(-2\pi j)e^{j\omega_p t} = -je^{j(\omega_{pr} - j\delta)t} = -je^{j\omega_{pr}t}e^{\delta t} \qquad \text{for } t < 0,$$

and $h(t) = 0$ for $t > 0$ where we take the contour C_1 for $t < 0$ and C_2 for $t > 0$ as shown in Fig. 7.4.2. This is noncausal.

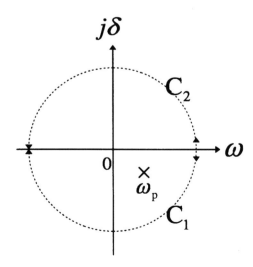

Figure 7.4.2 Contours for calculating the impulse response.

The zeros, however, are distributed in the entire complex frequency plane, while the poles are located on the upper half-plane only. We will investigate the distribution statistics of poles and zeros of acoustic systems using the general formulae with Green's functions.

7.4.2 Zeros and Residues

The input (driving point) impedance corresponds to the Green's function when the observation point is taken at the unit point source position,

$$G(\mathbf{x}, \mathbf{x}) = \sum_N \frac{|u_N(\mathbf{x})|^2}{\lambda_N - \lambda},$$

where $u_N(\mathbf{x})$ is Nth eigenfunction and λ_N is the Nth eigenvalue. As we already described in the previous chapter, the poles are distributed following a Poisson-like distribution. The occurrence of zeros occurring between an adjacent pair of poles depends on relationship between the signs of their residues (Lyon, 1984). We can verify the relationship between the zero occurrences and the residue sign conditions.

Lyon introduced a formula by considering a test frequency between two adjacent poles as

$$H(\omega) \equiv \frac{A}{\omega - \omega_1} + \frac{B}{\omega - \omega_2} + R = y_1 + y_2 + R,$$

where R is a remainder function corresponding to the sum of the contributions from poles excluding the two adjacent poles. If we approximate the remainder function R as a slowly varying function (almost constant in the interval between the pair of poles) of the test frequency ω, we can analyze the zero occurring conditions in the interval between the adjacent pair of poles.

Figure 7.4.3 is a schematic illustration of zero-occurrence conditions as shown by Lyon. A zero is produced between the two adjacent poles when

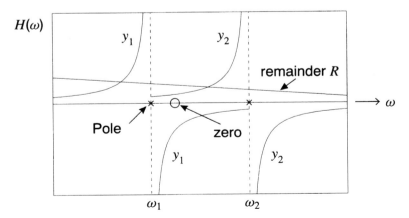

Figure 7.4.3 Zero occurring between two adjacent poles.

they have residues of the same sign. Such a zero is located on the line connecting the poles (pole-line). We call this type of zero an on-line zero.

The effect of the remainder function on the zero-occurrence process is only a slight change of the location of the zero. For simplicity, assume that the remainder function is negligibly small. The zero is the root of

$$H(\omega) = \frac{A}{\omega_z - \omega_1} + \frac{B}{\omega_z - \omega_2} = 0.$$

We can obtain the zero,

$$\omega_z = \omega_0 + \Delta\omega \frac{A - B}{A + B}$$

where

$$\omega_0 = \frac{\omega_1 + \omega_2}{2}, \qquad \Delta\omega = \frac{\omega_2 - \omega_1}{2},$$
$$\omega_2 = \omega_0 + \Delta\omega, \qquad \omega_1 = \omega_0 - \Delta\omega$$

and $\omega_2 > \omega_1$. Therefore, when both A and B have the same sign, we have the relationship

$$\left| \frac{A - B}{A + B} \right| < 1.$$

Thus we get an on-line zero which satisfies the relationship $\omega_1 < \omega_z < \omega_2$. When the pair of residues has the same sign, we have a zero in the interval of the pole pair.

The input impedance corresponding to

$$G(\mathbf{x}, \mathbf{x}) = \sum_N \frac{|u_N(\mathbf{x})|^2}{\lambda_N - \lambda}$$

has no sign changes between the adjacent pair of residues. Therefore, every interval of adjacent poles in the input impedance has an on-line zero.

When the pair of residues has opposite signs, the zero-occurrence process is rather complicated. If a pair of two adjacent poles has opposite sign residues, the occurrence of zeros depends on the remainder function R. Let us return to the approximate equation for the TF,

$$H(\omega) \equiv \frac{A}{\omega - \omega_1} + \frac{B}{\omega - \omega_2} + R = y_1 + y_2 + R.$$

The equation for obtaining the zeros is

$$H(\omega) \equiv \frac{A}{\omega_z - \omega_1} + \frac{B}{\omega_z - \omega_2} + R = 0.$$

We have to solve the quadratic equation to get the roots.

Change the variable. Setting the new variable

$$\Omega = \omega - \omega_0, \qquad \omega - \omega_1 = \Omega + \Delta\omega, \qquad \text{and } \omega - \omega_2 = \Omega - \Delta\omega,$$

we get

$$\frac{A}{\Omega + \Delta\omega} + \frac{B}{\Omega - \Delta\omega} + R = 0.$$

This equation can be rewritten in the quadratic equation form

$$R\Omega^2 + (A + B)\Omega - [R(\Delta\omega)^2 + (A - B)\Delta\omega] = 0.$$

Thus, we can obtain the solutions as

$$\omega_z = \omega_0 + \frac{-(A + B) \pm \sqrt{(A + B)^2 + 4R[R(\Delta\omega)^2 + (A - B)\Delta\omega]}}{2R}.$$

For simplicity, assume that the magnitudes of both residues are the same, that is, $|A| = |B| = A > 0$. If the residues have the same sign, we get zeros such as

$$\omega_{z1} \equiv \omega_0 + \sqrt{\frac{A^2}{R^2} + (\Delta\omega)^2 + \frac{-A}{R}}$$

and

$$\omega_{z2} \equiv \omega_0 - \sqrt{\frac{A^2}{R^2} + (\Delta\omega)^2 + \frac{-A}{R}}.$$

Consequently, as already described, we have one zero on the pole-line, when the residues have the same sign. This is because one of the zeros derived above is always located on the pole line between the pair of poles. If we have a remainder which satisfies the relationship $A/R > 0$, we get the on-line zero at $\omega = \omega_{z1}$ in the interval of the pair of poles. This can be reconfirmed by checking that

$$-\Delta\omega < \sqrt{\frac{A^2}{R^2} + (\Delta\omega)^2 + \frac{-A}{R}} < \Delta\omega$$

holds well. Conversely, when $A/R < 0$, we get the on-line zero at $\omega = \omega_{z2}$ in the interval.

If the residues have opposite signs, there are three cases for zeros: in one case there are no zeros between the pair of poles; another shows the double zeros on the pole-line; the third case involves a pair of zeros which are symmetrically located in upper and lower half-planes in the complex frequency plane where the test frequency is expanded into the complex plane. Figure 7.4.4 shows these three cases.

Let us assume that the residues have opposite signs and $|A| = |B| = A > 0$. The zeros are given as

$$\omega_z = \omega_0 + \frac{-(A+B) \pm \sqrt{(A+B)^2 + 4R[R(\Delta\omega)^2 + (A-B)\Delta\omega]}}{2R}$$

$$= \omega_0 \pm \Delta\omega \sqrt{1 + \frac{2A}{\Delta\omega R}}.$$

Consequently, we have double zeros when the remainder function is negative and satisfies the relationship

$$1 > 1 + \frac{2A}{\Delta\omega R} > 0.$$

We have the coincident double zeros when

$$1 + \frac{2A}{\Delta\omega R} = 0.$$

If the remainder function is positive, the zeros are outside the interval. There are no zeros in the interval in this case.

If the remainder function is negative and the relationship

$$1 + \frac{2A}{\Delta\omega R} < 0$$

holds well, then the zeros become complex. The complex roots are given by

$$\omega_0 \pm \Delta\omega \sqrt{1 + \frac{2A}{\Delta\omega R}} = \omega_0 \pm j\Delta\omega \sqrt{-\left(1 + \frac{2A}{\Delta\omega R}\right)}.$$

The zeros above are located symmetrically with respect to each other at equal distances from the pole-line. If we take the negative sign for the residue A, the conditions for the sign of the remainder function are inverted.

The residue sign conditions of Green's functions are complicated even for 1-D cases with a homogeneous boundary condition such that $u(x = 0) = u(x = L) = 0$. When we take a receiving position different from the source position, the residue signs change rather randomly. We will investigate the probability of residue sign change in the next section.

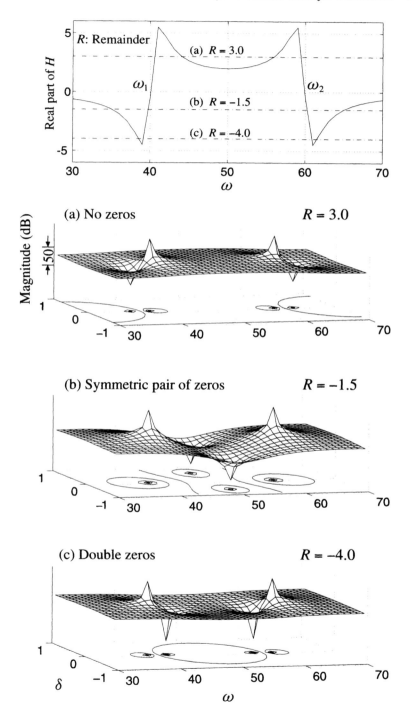

Figure 7.4.4 Occurrence of zeros under opposite-sign residue conditions.

7.4.3   Residue Sign Change Probabilities

Let us take an example of Green's function for a 2-D case as

$$H((x_S, y_S), (x_M, y_M))$$

$$\equiv \sum_{l,m} \frac{\sin\left(\dfrac{l\pi x_S}{L_x}\right) \sin\left(\dfrac{m\pi y_S}{L_y}\right) \sin\left(\dfrac{l\pi x_M}{L_x}\right) \sin\left(\dfrac{m\pi y_M}{L_y}\right)}{\omega^2 - \omega_{lm}^2}$$

where (x_S, y_S) is the unit point source location, (x_M, y_M) is the observation location. The lmth eigenfrequency is given by

$$\omega_{lm} = c\sqrt{\left(\frac{l\pi}{L_x}\right)^2 + \left(\frac{m\pi}{L_y}\right)^2}, \qquad (\text{rad s}^{-1})$$

where L_x and L_y are the lengths (m) of the sides of the rectangular boundary, and c is the sound speed (m s^{-1}) in the medium.

We can estimate the expected number of zeros using the probability in the residue sign changes, if we neglect the case for double zeros (Lyon, 1984). The residue signs depend on the source and receiver positions. Figure 7.4.5 shows the probability that an adjacent pair of poles has opposite sign residues, when we randomly sample the source and receiver positions in the 2-D rectangular space. The probability is linearly proportional to kr when kr is smaller than 2, while it does not significantly depend on the distance when kr is greater than 2. Here, k is the wavenumber of the test frequency and r is the distance between the

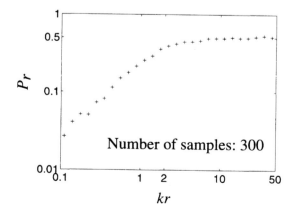

Figure 7.4.5 Probability of residue sign changes for 300 pairs of source and receiving positions taken randomly in a 2-D space where $L_x = 4.5$ and $L_y = 4.5 \times 2^{1/3}$.

source and receiving positions. Lyon (1984) investigated these probability characteristics theoretically and experimentally.

7.4.4   Distribution of Zeros and Transfer Function Residue Models

The expected number of zeros on the pole-line which we call on-line zeros can be approximately estimated by

$$\overline{N_{z,\,\text{on-line}}} = N_p(1 - p_r),$$

where N_p denotes the number of poles in a frequency interval of our interest and p_r denotes the probability of having residue sign changes. This probability is estimated from Fig. 7.4.5 and the equation above neglects the number of double zeros. The number of off-line zeros in the upper half-plane (or lower half-plane because of symmetry) can be estimated by

$$\overline{N_{z,\,\text{upper half}}} = \frac{N_p p_r}{2}.$$

Here we assume equal probability of having no on-line zeros or a pair of symmetric off-line zeros in an interval of opposite residue pole pairs.

As Lyon (1984) pointed out, we can see in Fig. 7.4.5 that as the distance between the source and receiving positions increases, the possibility of residue sign changes increases. Consequently, the number of on-line zeros decreases, while the number of off-line zeros increases as the distance becomes large. The probability p_r, however, does not significantly depend on kr when kr is greater than 2, as we can see in Fig. 7.4.5. Thus, when kr is greater than 2, the number of on-line zeros can be estimated by (Lyon, 1984)

$$\overline{N_{z,\,\text{on-line}}} = \frac{N_p}{2},$$

where $p_r = \frac{1}{2}$, and the number of off-line zeros in the upper half-plane (or lower half-plane because of symmetry) is given by

$$\overline{N_{z,\,\text{upper half}}} = \frac{N_p}{4}.$$

The zeros of the TFs in the complex frequency plane can be counted by an integral formula for the analytical function

$$\frac{-1}{2\pi j} \int_C \frac{H'(u)}{H(u)} du = N_p - N_z$$

where $u = \omega + j\delta$, and C denotes a closed curve by which our area of interest is surrounded in the complex plane, N_p is the number of poles located in the area, and N_z is the number of zeros in the area. We can calculate the number of zeros in the area surrounded by a closed curve if we know the number of poles in the area (Tohyama, Suzuki and Ando, 1995).

Figure 7.4.6 shows the averaged number of on-line and off-line zeros in the upper half-plane counted using random TF models as

$$H(\omega) \equiv \sum_{n=1}^{N} \frac{A_n}{\omega - \omega_n},$$

where $|A_n| = 1$. We vary the probabilities of residue sign changes between the adjacent pair of poles according to a binomial distribution. The poles are assumed to be distributed according to a Poisson distribution on the real frequency axis which has the average of N_p. We can verify that the number of on-line zeros follows the solid line in Fig. 7.4.6,

$$\overline{N_{z,\,\text{on-line}}} = N_p(1 - p_r),$$

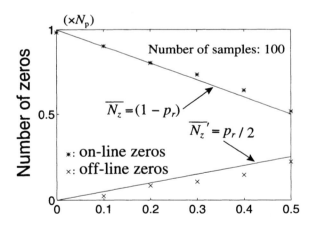

Figure 7.4.6　Distribution of zeros for random TFs whose poles are distributed according to a Poisson distribution and the residue signs are randomly changed according to a binomial distribution. N_p is the average number of poles in the frequency interval of interest.

while the number of off-line zeros can be estimated by

$$\overline{N_{z,\,\text{upper half}}} = \frac{N_p p_r}{2}.$$

7.5 Phase of Room Transfer Functions

7.5.1 Phase and Phase Velocity

The magnitude characteristics of TFs are related to the power spectra of transmitted signals through the systems of interest. The phase characteristics show the phase relationship between the input and output signals. The time delay between the input and output signals is given by

$$\tau_p \equiv \frac{-\phi}{\omega} \quad \text{(s)},$$

where ϕ (rad) denotes the TF phase, and ω is the angular frequency (rad s^{-1}). If we take a pair of source and receiving positions with a separation r (m) in an acoustic system such as a room; then

$$\tau_p \equiv \frac{-\phi}{\omega} = \frac{r}{c_p} \quad \text{(s)},$$

where c_p is the phase velocity defined by

$$c_p \equiv \frac{\omega}{k} \quad \text{(m s}^{-1}\text{)}$$

and k is the wavenumber. If c_p depends on the frequency, the medium of the system is called dispersive. An elastic plate is dispersive for bending waves. In nondispersive medium, the delay is independent of the frequency and the phase is linearly proportional to the frequency.

7.5.2 Group Delay

The group delay is a local property of the phase. The group delay between the signals is given by

$$\tau_g \equiv -\frac{\partial \phi}{\partial \omega} = r\frac{\partial k}{\partial \omega} = \frac{r}{c_g} \quad \text{(s)},$$

where c_g is the group velocity defined by

$$c_g \equiv \frac{\partial \omega}{\partial k} \quad \text{(s)}$$

and c_g gives an envelope delay of a compound signal.

Suppose that we have a linear system with a flat magnitude response and a phase response $\phi(\omega)$. If we take an input signal

$$x_{\text{in}}(t) \equiv \sin\left(\omega_c + \frac{\Delta\omega}{2}\right)t + \sin\left(\omega_c - \frac{\Delta\omega}{2}\right)t$$

$$= 2\sin\omega_c t \cos\Delta\omega t,$$

then the output signal is given by

$$x_{\text{out}}(t) \cong \sin\left\{\left(\omega_c + \frac{\Delta\omega}{2}\right)t + \left(\phi_c + \frac{\Delta\phi}{2}\right)\right\}$$

$$+ \sin\left\{\left(\omega_c - \frac{\Delta\omega}{2}\right)t + \left(\phi_c - \frac{\Delta\phi}{2}\right)\right\}$$

$$= 2\sin(\omega_c t + \phi_c)\cos(\Delta\omega t + \Delta\phi)$$

$$\cong 2\sin(\omega_c t + \phi_c)\cos\left(\Delta\omega t + \frac{\partial\phi}{\partial\omega}\Delta\omega\right)$$

$$= 2\sin(\omega_c t + \phi_c)\cos\{\Delta\omega(t - \tau_g)\}.$$

The group delay is the delay of the envelopes between the input and output signals as shown by Fig. 7.5.1. The group delay depends on the frequency even for a nondispersive medium because of the reflection waves.

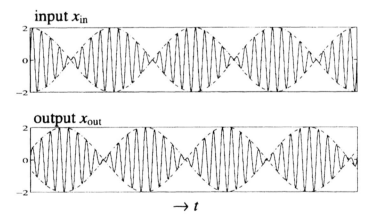

Figure 7.5.1   Delay between envelopes of two signals.

7.5.3   Accumulated Phase

The phase is estimated by the distribution of poles and zeros as well as the magnitude. The phase change of $-\pi$ occurs when the test frequency passes a pole in the upper half-plane or a zero located below the real frequency axis. However, $+\pi$ phase change occurs when passing a zero located above the real frequency axis. The phase lags due to the poles or zeros below the real frequency axis can be compensated by the zeros in the upper half-plane. Figure 7.5.2 illustrates an example of the phase changes due to the poles and zeros.

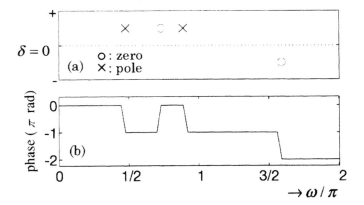

Figure 7.5.2 Phase accumulation due to poles and zeros.

The TF phase is defined as the argument of the TF which is represented by

$$H(\omega) = |H(\omega)| \exp(j\phi(\omega)).$$

If we take the imaginary part of natural log of the TF, the TF phase is

$$\phi(\omega) = \sum_{\text{for zeros } \omega_z} \arg(\omega - \omega_z) - \sum_{\text{for poles } \omega_p} \arg(\omega - \omega_p).$$

The phase is accumulated in proportion to the difference between the sum of the numbers of poles and zeros in the lower half-plane and the number of zeros in the upper half-plane including on the real frequency axis.

The accumulated phase can be approximately estimated by the relation (Lyon, 1983),

$$\phi(\omega) \cong -\pi(N_p + N_z^+ - N_z^- - N_{z,\text{on-line}}) \pm \frac{\pi}{2},$$

where N_p is the number of poles, N_z^+ denotes the number of zeros in the lower half-plane and N_z^- the zeros in the upper half-plane as shown in Fig. 7.5.3, and the last term arises from the possibility of a pole or zero near $\omega = 0$.

The phase for a 1-D TF such as

$$H(\omega) \equiv j\rho c \frac{\sin k(L - x)}{\cos kL}$$

is estimated (Lyon, 1984) as

$$\phi(k) \approx -\pi(N_p - N_z) = -\pi\left(\frac{k}{\pi/L} - \frac{k}{\pi/(L - x)}\right)$$

$$= -kx,$$

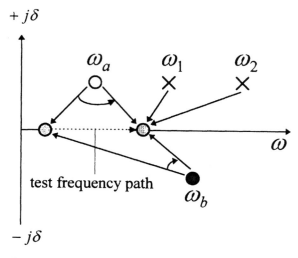

Figure 7.5.3   Change in phase of TF as test frequency moves along the real frequency axis.

where N_z denotes the number of on-line zeros below the wavenumber k of the test frequency, L (m) is the length of the system, and x (m) is the distance from the source which is located at one end. As we described in Chapter 6, kx denotes the phase change between the waves at two positions with a distance x. The progressive wave phase changes the phase by kx, while the wave travels a distance x. The TF for a 1-D system composed of eigenfunctions makes the phase change similar to that of a 1-D progressive wave in an infinite medium in spite of the reflections from both ends.

The accumulated phase for 2-D systems is different from the 1-D cases. The accumulated phase can be calculated from an integration formula as well as estimating the number of zeros,

$$\phi(\omega_2) - \phi(\omega_1) = \text{Im} \int_{\omega_1}^{\omega_2} \frac{H'(\omega)}{H(\omega)} \, d\omega,$$

where Im denotes taking the imaginary part and $H'(\omega)$ is the first derivative of $H(\omega)$ with respect to ω (van Eeghem *et al.*, 1996).

Figure 7.5.4 shows examples of the accumulated phase up to the test frequency ω using the random TF model

$$H(\omega) \equiv \sum_{n=1}^{N} \frac{A_n}{\omega - \omega_n}$$

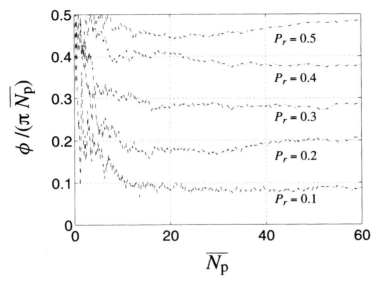

Figure 7.5.4 Phase trend for random TFs where the poles are distributed according to a Poisson distribution and residue signs change randomly according to a binomial distribution, and 100 TFs are used for calculation.

where $|A_n| = 1$. The poles are distributed according to a Poisson distribution and residue signs change randomly according to a binomial distribution. The solid line shows the estimation of

$$\phi(\omega) \cong -\pi(N_p + N_z{}^+ - N_z{}^- - N_z, \text{ on-line}) = -\pi(N_p - N_{z, \text{ on-line}})$$

$$= -\pi\{N_p - N_p(1 - p_r)\} = -\pi N_p p_r,$$

where p_r is the probability of residue sign changes and off-line zeros are symmetrically distributed with respect to the pole line. The phase accumulations by the pairs of off-line zeros are canceled out on the real frequency axis because of symmetry. We can estimate the maximum phase accumulation by setting $p_r = \frac{1}{2}$ (Lyon, 1984). That is,

$$\phi_{\text{Max}} = \frac{-N_p \pi}{2},$$

7.6 Summary

The statistical approach has been developed as a practical method for characterizing the TFs for complex acoustic systems. In this chapter, we described random sound fields composed of direct and reflection waves. The polynomial TF models which represent random sound fields can be characterized by the distribution of zeros, which are complex roots of the polynomials. The variances of the coefficients of random polynomials decrease exponentially as the polynomial order increases. This exponentially decreasing function can be calculated from the reverberation time, which is estimated using the mean free path of the reflections and the absorption coefficients.

The fine structures of the zero-occurrence process can be described on the basis of the conditions of residue sign change between the two adjacent poles. The distribution of zeros and phase characteristics can be analyzed by introducing the residue sign change probabilities into TF models using poles and their residues.

8 Deconvolution and Inverse Filters

8.0 Introduction

Sound field control used in sound image projection for virtual reality technology requires equalization of the transfer functions including both magnitude and phase of the acoustic systems. This equalization is called inverse filtering or deconvolution. Deconvolution is a key technology for dereverberation of reverberant speech signals or source waveform recovery for machine diagnostics (Lyon, 1987). These inversion problems are formidable, particularly concerning the stability of inverse filters. This chapter describes fundamentals of inverse filtering for acoustic TFs. Causality and stability of inverse filters, analytic signals, minimum-phase and all-pass decomposition, cepstrum, inverse filtering for minimum-phase components, and phase equalization are discussed. Most of these issues overlap Chapter 4; however, expressions in this chapter are based on continuous function forms with simulation examples. Inverse filtering for most practical situations using finite impulse response filters are described in Chapter 9.

8.1 Convolution and Integral Equations

The relationship between the input and output responses in a linear system is described as

$$y(t) = \int_0^t x(t - \tau)h(\tau)\, d\tau,$$

where $x(t)$ is an input signal which starts at $t = 0$, $h(t)$ is the impulse response, and $y(t)$ denotes the output response. This formula represents a convolution. The inverse problem is to find $x(t)$ which solves the integral equation when both the output signal $y(t)$ and the impulse response $h(t)$ are known. This process is called *deconvolution.*

Deconvolution can also be considered in the frequency domain. The integral equation stated above is rewritten as the product in the frequency domain as

$$Y(\omega) = X(\omega)H(\omega),$$

where $X(\omega)$, $H(\omega)$, and $Y(\omega)$ are Fourier transforms of $x(t)$, $h(t)$, and $y(t)$, respectively. The deconvolution is described as

$$X(\omega) = \frac{Y(\omega)}{H(\omega)} = Y(\omega)H^{-1}(\omega)$$

in the frequency plane. $H^{-1}(\omega)$ is called an *inverse filter*.

Figure 8.1.1 gives an example of an inverse filter. Figure 8.1.1a shows the impulse response of a system; Fig. 8.1.1b illustrates its inverse filter's impulse response; and Fig. 8.1.1c shows the convolution of the system and inverse system impulse responses. Inverse filtering makes the system's original impulse response to be an ideal impulse.

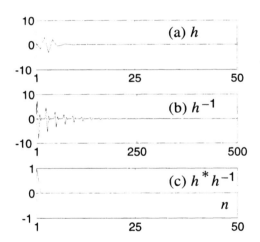

Figure 8.1.1 An example of inverse filter. (a) System impulse response. (b) Inverse filter's impulse response. (c) Convolution of system and inverse system impulse responses.

8.2   Causality of Inverse Filters

8.2.1   Causality

The inverse filter defined by $1/H(\omega)$ for the TF of $H(\omega)$, is not always causal. System causality depends on the locations of the poles, as was mentioned in Chapters 6 and 7. If we assume $\exp(j\omega t)$ time dependency for input and output sinusoidal signals, the poles of the TF between input and output signals are located in the upper half-plane in the complex frequency plane. When the poles are located in the lower half-plane, the system is noncausal.

8.2.2   Real and Imaginary Parts of the Causal System Transfer Function

As described in Chapter 4, the real and imaginary parts of the causal TFs are related to each other. Let us take the impulse response $h(t)$ of a real function and its Fourier transform as

$$H(f) = \int_{-\infty}^{+\infty} h(t)e^{-j2\pi ft}\, dt \equiv H_r(f) + jH_i(f),$$

where

$$H_r(f) \equiv \int_{-\infty}^{+\infty} h(t)\cos(2\pi ft)\, dt$$

and

$$H_i(f) \equiv -\int_{-\infty}^{+\infty} h(t)\sin(2\pi ft)\, dt.$$

The real part $H_r(f)$ is an even function of f, while the imaginary part $H_i(f)$ is an odd function.

If we take the inverse Fourier transform of $H(f)$, we get

$$h(t) = \int_{\infty}^{+\infty} H(f)e^{j2\pi ft}\, df \equiv h_r(t) + h_i(t),$$

where

$$h_r(t) \equiv \int_{-\infty}^{+\infty} H_r(f)e^{j2\pi ft}\, df = 2\int_{0}^{+\infty} H_r(f)\cos(2\pi ft)\, df$$

and

$$h_i(t) \equiv \int_{-\infty}^{+\infty} jH_i(f)e^{j2\pi ft}\, df = -2\int_{0}^{+\infty} H_i(f)\sin(2\pi ft)\, df.$$

Here $h_r(t)$ is an even function of t and $h_i(t)$ is an odd function of t.

If the impulse response $h(t)$ is causal so that $h(t) = h(t)$ at $t > 0$ and $h(t) = 0$ at $t \leq 0$, then, as shown in Fig. 8.2.1, we get the relationships

$$h_r(t) = h_i(t) \qquad \text{at } t > 0,$$
$$h_r(t) = h_i(t) = 0 \qquad \text{at } t = 0,$$

and

$$h_r(t) = -h_i(t) \qquad \text{at } t < 0.$$

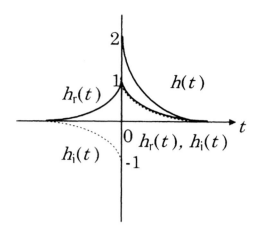

Figure 8.2.1   Decomposition of a causal function into even $h_r(t)$ and odd $h_i(t)$ functions.

Consequently, the real part of the TF,

$$H_r(f) \equiv \int_{-\infty}^{+\infty} h(t)\cos(2\pi f t)\,dt$$

$$= \int_{-\infty}^{+\infty} [h_r(t) + h_i(t)]\cos(2\pi f t)\,dt$$

$$= \int_{-\infty}^{+\infty} h_r(t)\cos(2\pi f t)\,dt$$

$$= 2\int_{0}^{+\infty} h_r(t)\cos(2\pi f t)\,dt$$

$$= 2\int_{0}^{+\infty} h_i(t)\cos(2\pi f t)\,dt$$

$$= -4\int_{0}^{+\infty}\cos(2\pi f t)\,dt\left[\int_{0}^{+\infty} H_i(f')\sin(2\pi f' t)\,df'\right]$$

and similarly the imaginary part becomes

$$H_i(f) \equiv -\int_{-\infty}^{+\infty} h(t)\sin(2\pi f t)\,dt = -2\int_{0}^{+\infty} h_i(t)\sin(2\pi f t)\,dt$$

$$= -2\int_{0}^{+\infty} h_r(t)\sin(2\pi f t)\,dt$$

$$= -4\int_{0}^{+\infty}\sin(2\pi f t)\,dt\left[\int_{0}^{+\infty} H_r(f')\cos(2\pi f' t)\,df'\right].$$

Thus the real and imaginary parts for the causal TF are converted into each other. This relation is called the Hilbert transform.

8.2.3 Analytic Signal

This section is not directly related to the inverse filters and causal systems. However, if we apply the concept of causality to the frequency domain, we can get an alternative formula for time signal representation which we call an analytic signal. Let us take an example such as

$$\cos \omega t = \frac{e^{j\omega t} + e^{-j\omega t}}{2}.$$

If we add an imaginary part to a real signal, then we can represent the 'complex signal' using only positive frequency components such as

$$\cos \omega t + j \sin \omega t = e^{j\omega t}.$$

This type of complex signal is called an analytic signal.

Suppose that we have a real signal $x(t)$ and its Fourier transform

$$X(f) \equiv X_r(f) + jX_i(f)$$

where $X_r(f)$ and $X_i(f)$ are the real and imaginary parts, respectively. If we add an imaginary component to the real signal such as

$$z(t) \equiv x(t) + jy(t),$$

then the Fourier transform of the complex signal becomes

$$Z(f) = X_r(f) + jX_i(f) + j[Y_r(f) + jY_i(f)]$$
$$= [X_r(f) - Y_i(f)] + j[X_i(f) + Y_r(f)]$$

where $Y(f)$ denotes the Fourier transform of $y(t)$ and

$$Y(f) \equiv Y_r(f) + jY_i(f).$$

Here we require the real and imaginary parts of $Y(f)$, $Y_r(f)$ and $Y_i(f)$, to satisfy the relationships

$$Y_i(f) = X_r(f), \qquad Y_r(f) = -X_i(f) \qquad \text{in } f \leq 0$$

and

$$Y_i(f) = -X_r(f), \qquad Y_r(f) = X_i(f) \qquad \text{in } f > 0$$

in order to make $Z(f)$ causal. Therefore,

$$Z(f) = 0 \quad \text{in } f \leq 0 \quad \text{and} \quad Z(f) = 2X(f) \quad \text{in } f > 0.$$

Figures 8.2.2 to 8.2.5 show examples of analytic signals. Figure 8.2.2a is the original real signal $x(t)$; Figs 8.2.2b and 8.2.2c illustrate the real and imaginary parts of the Fourier transform of $x(t)$. Figure 8.2.3a is the imaginary part $y(t)$ to be added to the original real signal $x(t)$; Figs 8.2.3b and 8.2.3c show the real and imaginary parts of the Fourier transform of $y(t)$. Figure 8.2.4a shows the real part of Fourier transform of $z(t)$, and similarly Fig. 8.2.4b is the imaginary part. We can see the 'causality' in the frequency domain and $Z(f) = 2X(f)$ in $f > 0$.

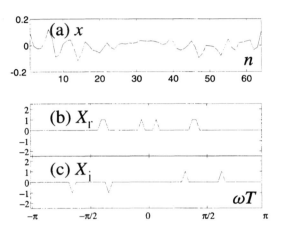

Figure 8.2.2 A time waveform and its Fourier transform: (a) time signal; (b) real part of Fourier transform of the signal; (c) the imaginary part of the Fourier transform.

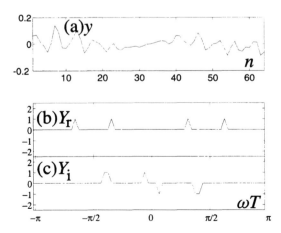

Figure 8.2.3 Imaginary signal to be added into the original real signal: (a) time signal; (b) real part of Fourier transform of the signal in (a); (c) the imaginary part of the Fourier transform.

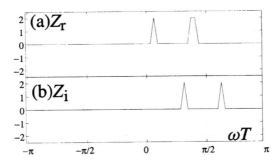

Figure 8.2.4 Analytic signal: (a) the real part of Fourier transform of the analytic signal; (b) the imaginary part of the Fourier transform.

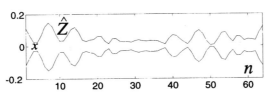

Figure 8.2.5   Envelope.

Figure 8.2.5 demonstrates the envelope $\hat{z}(t)$ of the original signal which is defined as

$$\hat{z}(t) \equiv \pm|z| = \pm\sqrt{x^2(t) + y^2(t)} \quad \text{and} \quad \hat{\theta}(t) \equiv \arg(z) = \tan^{-1}\frac{y(t)}{x(t)}.$$

8.2.4   Causality of Inverse Filters

The poles of a TF make the zeros of its inverse filter. The zeros of the original TF are the poles of the inverse filter. Thus, the causality of inverse filters depends on the zero locations of the TFs. The zeros in the upper half-plane become the 'causal poles' of the inverse filter, while the zeros in the lower half-plane are the 'noncausal poles' of the inverse filter. Zeros of the inverse filters are always in the upper half-plane, since stable TFs have no poles in the lower half-plane.

8.3   Inverse Filters for Minimum-phase Components

8.3.1   Minimum-phase and All-pass Components of the Transfer Function

As described in Chapter 4, the TF can be written in product form as

$$H(\omega) = H_{\min}(\omega)H_{ap}(\omega)$$

on the real frequency axis in the complex frequency plane. Here $H_{min}(\omega)$ and $H_{ap}(\omega)$ are the minimum-phase and all-pass components respectively, and they satisfy the relations

$$|H(\omega)| = |H_{min}(\omega)| \qquad \text{and} \qquad |H_{ap}(\omega)| = 1.$$

Suppose that the TF is written in factorized form as

$$G(k, \mathbf{x}, \mathbf{x}') = \sum_N \frac{u_N(\mathbf{x})u_N^*(\mathbf{x}')}{k_N^2 - k^2} \equiv \sum_N \frac{a_N(\mathbf{x}, \mathbf{x}')}{k_N^2 - k^2}$$

$$\equiv \frac{N(\omega, \mathbf{x}, \mathbf{x}')}{D(\omega, \mathbf{x}, \mathbf{x}')} = \frac{(\omega - \omega_{z1})\ldots(\omega - \omega_{zn})}{(\omega - \omega_{p1})\ldots(\omega - \omega_{pm})} \equiv H(\omega),$$

where the poles ω_p are distributed in the upper half-plane and some of the zeros ω_z are located in the lower half-plane and others in the upper half-plane.

A model of pole–zero patterns for the minimum-phase and all-pass components is shown in Fig. 8.3.1. The zeros are symmetrically moved from below the real frequency axis to above the axis in Fig. 8.3.1b. The TF component constructed from these locations (Fig. 8.3.1b) of poles and zeros is called the minimum-phase component. In Fig. 8.3.1c 'new' poles are introduced in order to cancel the effects of zeros which are newly entered into the upper half-plane. The zeros in Fig. 8.3.1c are assigned to the locations of the original zeros in the lower half-plane. The poles

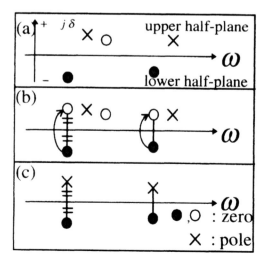

Figure 8.3.1 Examples of TF pole and zero locations. (a) An example of TF pole and zero locations. (b) Pole–zero arrangements for the minimum-phase part of the TF. (c) Locations of the poles and zeros for the all-pass part.

and zeros in Fig. 8.3.1c are located symmetrically with respect to the real frequency axis. The symmetric distribution of poles and zeros make the magnitude of the TF unity. This is because a symmetric pair of pole and zero locations is a complex conjugate pair and

$$\left| \frac{(\omega - \omega_0^*)}{(\omega - \omega_0)} \right| = 1$$

where * indicates taking the complex conjugate. Thus, we refer to the TF component shown by Fig. 8.3.1c as the all-pass component.

The magnitude of the minimum-phase component is equal to that of the TF itself. The notion of 'minimum phase' can be understood if we recall the phase changes due to the poles and zeros as described in Chapters 4–7. The zeros above the real frequency axis compensate the phase lags, while the poles and the zeros below the real frequency axis, produce the phase lags. Therefore, if the TF has no zeros below the real frequency axis, the total phase lag becomes minimum in the TFs, all of which have the same magnitudes and the same numbers of poles and zeros. Thus, we say that the TF is of minimum phase, when all the zeros are located in the upper half-plane. We can see no zeros in the lower half-plane in Fig. 8.3.1b, while all the zeros are located in the lower half-plane for the all-pass component, as shown in Fig. 8.3.1c.

8.3.2 Minimum-phase and All-pass Cepstrum

A complex function G such as a TF can be expressed as

$$G \equiv G_r + jG_i = |G|e^{j\theta},$$

where G_r and G_i denote the real and imaginary parts of G,

$$|G| = \sqrt{G_r^2 + G_i^2} \qquad \text{and} \qquad \tan^{-1} \frac{G_i}{G_r} = \theta.$$

Taking the natural logarithm of G,

$$\ln G = \ln |G| + j\theta.$$

Thus, the logarithmic magnitude and the phase can be split by taking the natural logarithm. We can extend the natural logarithm for a negative number into the complex number. For example,

$$\ln(-1) = \ln(e^{-j\pi}) = -j\pi,$$

where we take the argument of -1 in the principal value between $-\pi \leq \theta < \pi$.

Introducing the cepstrum, we can describe the relationship between the magnitude and the phase of the TF of a minimum-phase system as well as the real and imaginary parts of a causal impulse response. The phase and magnitude of the minimum-phase components were already mentioned in Chapter 4. Let us take a natural logarithm of the TF, H, as

$$\ln H \equiv \ln |H| + j\theta$$

where the real part denotes the log-magnitude and the imaginary part is the phase. As described in Section 8.2.2, the real and imaginary parts of the TF for a causal system can be converted into each other. Therefore, if we take the Fourier inverse transform of $\ln H$ and we can get a causal function, then the log-magnitude and the phase can be converted to each other.

The inverse Fourier transform of $\ln H$ is called the complex cepstrum. The complex cepstrum is written as

$$Cep_H(\tau) \equiv F^{-1}[\ln H(\omega)]$$

where F^{-1} denotes taking the inverse Fourier transform. We normally set $Cep_H(0) = 0$. The complex cepstrum is real when the impulse response of the system of interest is real. This is because

$$Cep_H(\tau) \equiv F^{-1}[\ln H(\omega)] = F^{-1}[\ln |H| + j\theta],$$

where

$$F^{-1}[\ln |H|] \equiv \int_{-\infty}^{+\infty} \ln |H| e^{j2\pi ft}\, df = 2 \int_0^{+\infty} \ln |H| \cos(2\pi ft)\, df$$

and

$$F^{-1}[j\theta] = \int_{-\infty}^{+\infty} j\theta(f) e^{j2\pi ft}\, df = -2 \int_0^{+\infty} \theta(f) \sin(2\pi ft)\, df.$$

The magnitude $|H| = \sqrt{H_r^2 + H_i^2}$ and the log magnitude $\ln |H|$ are even functions, while the phase $\tan^{-1}(H_i/H_r) = \theta$ is an odd function. Therefore the complex cepstrum becomes real, expressed as

$$Cep_H(\tau) = F^{-1}[\ln |H| + j\theta] = F^{-1}[\ln |H|] + F^{-1}[j\theta]$$

$$= 2 \int_0^{+\infty} \ln |H| \cos(2\pi ft)\, df - 2 \int_0^{+\infty} \theta(f) \sin(2\pi ft)\, df.$$

The complex cepstrum is formulated as the integral along the frequency axis. The integral can be evaluated by the singularities of the integrand. The singularities for the log magnitude of the TF are both the zeros and the poles of the TF, since a natural logarithm has singularities on those points. The poles of the TF are located in the upper half-plane, while the zeros, in general, are distributed throughout the whole complex plane. However, if we take the minimum-phase TF, which has no zeros in the lower half-plane, the singularities for the cepstrum are located only in the upper half-plane. Thus, the cepstrum for the minimum-phase TF is causal, since the integrand of the inverse Fourier transform has no singularities in the lower half-plane.

If the cepstrum is causal, then we can get the log-magnitude from the phase and similarly the phase can be obtained from the log-magnitude. Let us take the TF

$$H(\omega) = H_{\min}(\omega)H_{\text{ap}}(\omega)$$

on the real frequency axis in the complex frequency plane. If we take the cepstrum for the TF, the cepstrum is composed of the minimum-phase and all-pass cepstra. That is,

$$Cep_{\text{H}}(\tau) \equiv F^{-1}[\ln H(\omega)] = F^{-1}[\ln H_{\min}(\omega)] + F^{-1}[\ln H_{\text{ap}}(\omega)]$$

$$\equiv Cep_{\text{Hmin}}(\tau) + Cep_{\text{Hap}}(\tau),$$

where

$$Cep_{\text{Hmin}}(\tau) \equiv F^{-1}[\ln H_{\text{Hmin}}(\omega)],$$

$$Cep_{\text{Hap}}(\tau) \equiv F^{-1}[\ln H_{\text{ap}}(\omega)],$$

and $F^{-1}[*]$ denotes taking the inverse Fourier transform of $[*]$.

The complex cepstrum can be decomposed into the magnitude cepstrum, which is an even function, and the phase cepstrum, which is an odd function. For example, the complex cepstrum of a complex function, G can be decomposed as

$$Cep_{\text{G}}(\tau) \equiv F^{-1}(\ln |G|) + F^{-1}(j\theta) \equiv Cep_{\text{Gmag}}(\tau) + Cep_{\text{Gphase}}(\tau).$$

Here we set

$$Cep_{\text{Gmag}}(\tau) \equiv F^{-1}(\ln |G|),$$

$$Cep_{\text{Gphase}}(\tau) \equiv F^{-1}(j\theta),$$

and

$$G = |G|e^{j\theta}.$$

Thus, we get the complex cepstrum decomposition for the TF as follows:

$$Cep_{\text{H}}(\tau) \equiv F^{-1}[\ln H(\omega)] = F^{-1}[\ln H_{\min}(\omega)] + F^{-1}[\ln H_{\text{ap}}(\omega)]$$

$$= Cep_{\text{Hmin}}(\tau) + Cep_{\text{Hap}}(\tau)$$

$$= Cep_{\text{Hmin}-\text{mag}}(\tau) + Cep_{\text{Hmin}-\text{phase}}(\tau)$$
$$+ Cep_{\text{Hap}-\text{mag}}(\tau) + Cep_{\text{Hap}-\text{phase}}(\tau)$$
$$\equiv Cep_{\text{Hmag}}(\tau) + Cep_{\text{Hphase}}(\tau),$$

where

$$Cep_{\text{Hmag}}(\tau) \equiv Cep_{\text{Hmin}-\text{mag}}(\tau) + Cep_{\text{Hap}-\text{mag}}(\tau)$$
$$= Cep_{\text{Hmin}-\text{mag}}(\tau),$$
$$Cep_{\text{Hphase}}(\tau) \equiv Cep_{\text{Hmin}-\text{phase}}(\tau) + Cep_{\text{Hap}-\text{phase}}(\tau),$$
$$Cep_{\text{Hap}-\text{mag}}(\tau) = 0,$$

since

$$|H_{\text{ap}}| = 1 \qquad \text{and} \qquad |H_{\text{min}}| = |H|.$$

The minimum-phase cepstrum has no negative time component, since it is causal. As for a non-minimum-phase system, it can have poles in the upper half-plane and some or all of the zeros are in the lower half-plane. Therefore, we have singularities in both the lower and upper half-planes after taking the natural logarithm. Consequently, the cepstrum is noncausal for the non-minimum-phase systems. There are negative time components.

8.3.3   Minimum-phase and All-pass Decomposition

Minimum-phase and all-pass decomposition can be made using the complex cepstrum. The negative time components of the complex cepstrum come from the all-pass cepstrum, since no negative time components are included in the minimum-phase part. The all-pass cepstrum has only the phase cepstrum of an odd function. Figure 8.3.2 illustrates a TF cepstrum decomposition example. Figure 8.3.2a shows the complex cepstrum. If we subtract the odd function of all-pass phase cepstrum (Fig. 8.3.2b) from the complex cepstrum (Fig. 8.3.2a), then we get the minimum-phase cepstrum as shown by Fig. 8.3.2c.

The minimum-phase cepstrum again is decomposed into the magnitude and phase cepstra as illustrated by Fig. 8.3.3. The causal function can be written as the sum of even and odd functions. The even function for the causal minimum-phase cepstrum (Fig. 8.3.3a) is the magnitude cepstrum (Fig. 8.3.3b), and the odd function corresponds to the minimum-phase phase cepstrum (Fig. 8.3.3c).

The magnitude and phase of the minimum-phase TF are not independent of each other. We have no negative time components for the minimum-phase cepstrum. Therefore, we have the following relation:

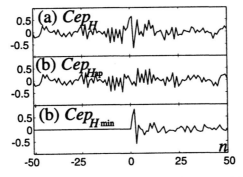

Figure 8.3.2 An example of TF cepstrum decomposition: (a) complex cepstrum; (b) all-pass cepstrum; (c) minimum-phase cepstrum.

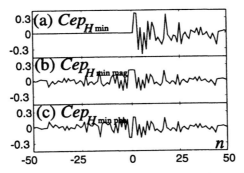

Figure 8.3.3 Minimum-phase cepstrum decomposition: (a) minimum-phase cepstrum; (b) minimum-phase magnitude cepstrum; (c) minimum-phase phase cepstrum.

$$Cep_{\text{Hmin-mag}}(|\tau|) = Cep_{\text{Hmag}}(|\tau|) = \begin{cases} -Cep_{\text{Hmin-phase}}(|\tau|) & \tau \leq 0 \\ Cep_{\text{Hmin-phase}}(|\tau|) & \tau > 0, \end{cases}$$

where we set $Cep(0) = 0$. We can construct the phase components from the magnitude, and conversely we can get the magnitude components from the phase cepstrum.

Figures 8.3.4 to 8.3.6 demonstrate examples of the decomposition of a TF. Figure 8.3.4 shows the impulse responses of the original TF (a), minimum-phase (b), and all-pass components (c). We can see that the minimum-phase impulse response record is shorter than those of the original and all-pass components. Figure 8.3.5 illustrates the magnitudes of the TF given by Fourier transform of the impulse responses shown by Fig. 8.3.4. The magnitude of the minimum-phase component is equal to the TF magnitude itself and the all-pass component magnitude is unity. Figure 8.3.6 similarly shows the TF-phase responses where the linear phase components are eliminated.

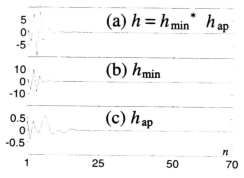

Figure 8.3.4 Impulse response decomposition example: (a) original impulse response; (b) minimum-phase impulse response; (c) all-pass response.

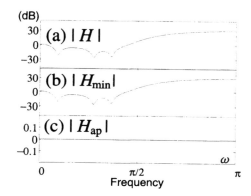

Figure 8.3.5 Decomposition of TF magnitude: (a) original TF magnitude; (b) minimum-phase magnitude ; (c) all-pass magnitude.

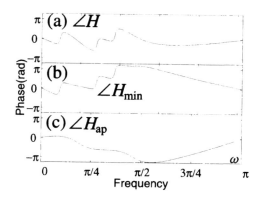

Figure 8.3.6 TF phase decomposition excluding the linear phase: (a) original TF phase; (b) minimum-phase phase; (c) all-pass phase.

8.3.4   Inverse Filter for the Minimum-phase Transfer Function

The TF is invertible when it is of minimum-phase, since there are no zeros in the lower half-plane. Inverse filters are always of minimum-phase, since all the zeros which are produced by the poles of the original TF are located in the upper half-plane. If we apply the inverse filter for the minimum-phase component of the TF to the original TF, we can get the all-pass response whose magnitude is unity for the output response,

$$H_{\min}^{-1}(\omega)H(\omega) = H_{\min}^{-1}(\omega)H_{\min}(\omega)H_{\mathrm{ap}}(\omega) = H_{\mathrm{ap}}(\omega).$$

Thus, the inverse filter for the minimum-phase component is a power spectrum equalizer.

Figures 8.3.7 and 8.3.8 show examples of the inverse filtering for the minimum-phase components of the TFs. We take the same TF example as shown in Fig. 8.3.4. The original impulse response in Fig. 8.3.7a is of non-minimum-phase. The minimum-phase and all-pass components are illustrated in Figs. 8.3.7b and 8.3.7c. Let us take an inverse filter for the minimum-phase components. The inverse filter's impulse response is shown in Fig. 8.3.8a. If we apply this inverse filter to the non-minimum-phase system whose impulse response is shown in Fig. 8.3.7a, then we can get the output response which is equal to the all-pass component of the system TF, as shown in Fig. 8.3.8b and 8.3.8c.

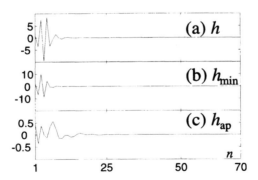

Figure 8.3.7   An example of non-minimum-phase system impulse response: (a) impulse response; (b) minimum-phase component; (c) non-minimum-phase component (all-pass component).

As we also see in Figs. 8.3.9 and 8.3.10, the TF magnitude is equalized to unity as shown in Fig. 8.3.9c, but the phase response is not compensated as illustrated in Fig. 8.3.10c. If the system is of minimum-phase without the all-pass part, the inverse filtering of the TF is possible.

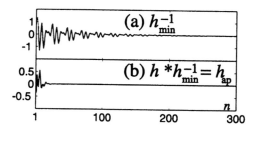

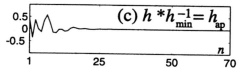

Figure 8.3.8 Inverse filtering for the minimum-phase component: (a) inverse filter's impulse response; (b) output response by inverse filtering; (c) close-up of (b).

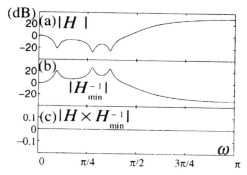

Figure 8.3.9 TF magnitude equalization by minimum-phase inverse filtering: (a) TF magnitude; (b) inverse filter's magnitude; (c) magnitude response equalized by the inverse filter.

8.3.5   Sound Image Projection System

We can apply the inverse filtering of minimum-phase components to a sound image projection system. Figure 8.3.11 is a diagram of a 2-channel sound image projection system. The filters required for the sound image projection at the phantom source position satisfy the following simultaneous equations:

$$Z_L(\omega) = X(\omega)G_L(\omega) + Y(\omega)H_L(\omega)$$

and

$$Z_R(\omega) = X(\omega)G_R(\omega) + Y(\omega)H_R(\omega),$$

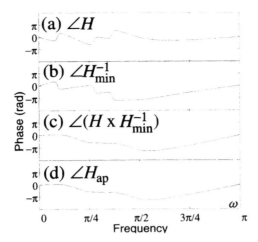

Figure 8.3.10 TF phase response excluding the linear phase and minimum-phase inverse filtering: (a) TF phase; (b) inverse filter's phase; (c) output phase response through the minimum-phase inverse filtering; (d) all-pass phase.

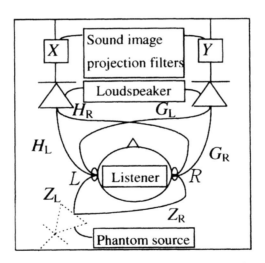

Figure 8.3.11 Two-channel sound image projection system

where Z_L (Z_R) denotes the transfer function between the phantom source and left (right) outer ear entrance at the listener's head position; G and H are HRTFs for the left- and right-channel loudspeakers, respectively. HRTF stands for the head related transfer function from a loudspeaker to a listener's outer ear entrance.

X and Y in Fig. 8.3.11 are filters for the sound image projection. These filters are obtained using the solutions of the simultaneous equations above as

$$X(\omega) = \frac{Z_L(\omega)H_R(\omega) - Z_R(\omega)H_L(\omega)}{H_L(\omega)G_R(\omega) - H_R(\omega)G_L(\omega)} \equiv \frac{X_N(\omega)}{D(\omega)}$$

and

$$Y(\omega) = \frac{Z_L(\omega)G_R(\omega) - Z_R(\omega)G_L(\omega)}{H_L(\omega)G_R(\omega) - H_R(\omega)G_L(\omega)} \equiv \frac{Y_N(\omega)}{D(\omega)}$$

where

$$D(\omega) \equiv H_L(\omega)G_R(\omega) - H_R(\omega)G_L(\omega).$$

These filters, however, are not causal when the denominator $D(\omega)$ is of non-minimum-phase.

$D(\omega)$ is written as the product of the minimum-phase and all-pass components:

$$D(\omega) = D_{\min}(\omega)D_{\mathrm{ap}}(\omega).$$

Let us take only the minimum-phase component as the denominators of the TFs for the filters, which are given by

$$\hat{X}(\omega) \equiv \frac{X_N(\omega)}{D_{\min}(\omega)} \quad \text{and} \quad \hat{Y}(\omega) \equiv \frac{Y_N(\omega)}{D_{\min}(\omega)}.$$

These causal filters satisfy

$$\hat{Z}_L(\omega) \equiv D_{\mathrm{ap}}Z_L(\omega) = \hat{X}(\omega)G_L(\omega) + \hat{Y}(\omega)H_L(\omega)$$

and

$$\hat{Z}_R(\omega) \equiv D_{\mathrm{ap}}Z_R(\omega) = \hat{X}(\omega)G_R(\omega) + \hat{Y}(\omega)H_R(\omega).$$

Binaural information such as the binaural level and phase differences are important for the sound image projection. The set of equations above shows that the binaural information is preserved without any deformation, even if we take only the minimum-phase components of the denominators to get the causal filters. The only information which is distorted by disregarding the all-pass parts of the denominators is monaural phase information. The filters preserve the binaural information that we want to reproduce.

8.3.6   Minimum-phase Waveform Recovery

Deconvolution is a key technology for source waveform recovery or dereverberation of speech. If a source waveform is of minimum-phase, then it can be recovered from the minimum-phase component of the output signal using the inverse filter constructed from the minimum-phase component of the TF.

Let us take again the same example of a non-minimum-phase system whose impulse response is shown in Fig. 8.3.12. Figure 8.3.13a is a minimum-phase source waveform. We can recover the minimum-phase source waveform from the minimum-phase component (Fig. 8.3.13c) of the reverberant response (Fig. 8.3.13b) through the system, even if the TF is of non-minimum-phase. Figure 8.3.14 shows the deconvolution example. Figure 8.3.14a is the inverse filter's impulse response for the minimum-phase component (Fig. 8.3.12b) and Fig. 8.3.14b is the recovered source waveform by deconvolution from the minimum-phase component of the reverberant signal (Fig. 8.3.13c) using the inverse filter. The minimum-phase waveform can be recovered.

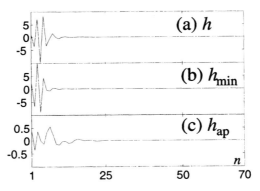

Figure 8.3.12   An example of non-minimum-phase system impulse response: (a) impulse response; (b) minimum-phase component; (c) all-pass component.

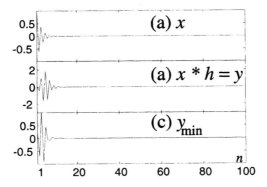

Figure 8.3.13   Minimum-phase source waveform and output response: (a) source waveform; (b) output reverberant response; (c) minimum-phase component of output response.

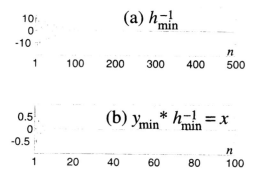

Figure 8.3.14   Inverse filtering and recovered source waveform: (a) impulse response of the inverse filter; (b) recovered source waveform.

Deconvolution using inverse filters is not applicable to practical situations where the TFs are unknown. Deconvolution when the TFs are unknown is called blind deconvolution. Blind deconvolution is an important technology for acoustic signal processing such as speech enhancement, but it is beyond the scope of this book.

8.4   Summary

This chapter explained the fundamentals of cepstrum processing and inverse filtering. The issues overlap Chapter 4; however, continuous function forms are used for expressions with numerical examples. Inverse filtering is magnitude and phase equalization for the TF which describes signal transmission in an acoustic system. Thus, inverse filtering is an important technology in acoustic signal processing such as sound image projection and waveform recovery. The TF decomposition into the minimum-phase and all-pass components is important for producing an inverse filter, since only the minimum-phase TF has a causal inverse filter. Taking the minimum-phase components, we can get a causal filter for a sound image projection system which preserves all the binaural information. The only information which is lost by disregarding the all-pass part is the monaural phase information. Blind deconvolution is important for applications such as speech dereverberation, since the TF is generally unknown, but it is formidable and advanced technologies beyond the scope of this book are required.

9 Linear Equations, Inverse Filters, and Signal Analysis

9.0 Introduction

In Chapter 8 we described inverse filtering based on transfer function models of acoustic systems. However, system representation which is employed in signal processing is practically based on discrete models. Although the discrete representation is obtained following the continuous models explained in previous chapters, the formulation for the discrete models is different. We also described the fundamentals of discrete signal processing methods in Chapters 2 to 5. This chapter describes inverse filtering and signal analysis based on the discrete models. These problems can be formulated using linear simultaneous equations. Inverse filtering is equivalent to getting an approximate solution for a set of simultaneous equations where the number of equations is higher than that of the unknowns. To obtain approximate solutions we use the normal equation which is employed in the linear regression of statistics, although our simultaneous equations basically are deterministic. The least square error method, signal representation by generalized discrete Fourier transforms, principal component analysis, and a pseudo-inverse method will be discussed.

9.1 Convolution

9.1.1 Impulse Response Matrix

Suppose that we have a discrete system whose impulse response is given by a sequence of $n + 1$ terms. The convolution is written as

$$y(n) = \sum_{k=0}^{n} x(n - k)h(k),$$

where x denotes the input signal sequence of $n + 1$ terms, h is the impulse response sequence, and y is the output signal sequence. The convolution is a linear transformation by the matrix H and can be formulated by a set of simultaneous linear equations as

$$y(0) = x(0)h(0)$$

$$y(1) = x(0)h(1) + x(1)h(0)$$

$$y(2) = x(0)h(2) + x(1)h(1) \qquad + x(2)h(0)$$

$$\vdots$$

$$y(n) = x(0)h(n) + x(1)h(n-1) + x(2)h(n-2) + \cdots + x(n)h(0)$$

$$y(n+1) = \qquad\qquad x(1)h(n) \qquad + x(2)h(n-1) + \cdots + x(n)h(1)$$

$$\vdots$$

$$y(2n) = \qquad\qquad\qquad\qquad\qquad\qquad\qquad x(n)h(n)$$

and equivalently in a matrix form as

$$\mathbf{y} \equiv H\mathbf{x},$$

where

$$\mathbf{y} \equiv \begin{pmatrix} y(0) \\ y(1) \\ \vdots \\ y(n) \\ y(n+1) \\ \vdots \\ y(2n) \end{pmatrix}, \qquad \mathbf{x} \equiv \begin{pmatrix} x(0) \\ \vdots \\ x(n) \end{pmatrix},$$

and

Here we call the matrix H an impulse response matrix. The output signal sequence is composed of $2n + 1$ terms. The number of terms resulting from the convolution of two sequences is $N + M - 1$, when the numbers of terms of the two sequences are N and M, respectively.

9.1.2 Data Matrix

The convolution can be rewritten as

$$y(0) = h(0)x(0)$$
$$y(1) = h(0)x(1) + h(1)x(0)$$
$$y(2) = h(0)x(2) + h(1)x(1) \qquad + h(2)x(0)$$
$$\vdots$$
$$y(n) = h(0)x(n) + h(1)x(n-1) + h(2)x(n-2)$$
$$\qquad\qquad\qquad\qquad\qquad + \qquad \cdots \qquad\qquad + h(n)x(0)$$
$$y(n+1) = \qquad\qquad h(1)x(n) \qquad + h(2)x(n-1) + h(3)x(n-2)$$
$$\qquad\qquad\qquad\qquad\qquad\qquad\qquad + \cdots + h(n)x(1)$$
$$\vdots$$
$$y(2n) = \qquad\qquad\qquad\qquad\qquad\qquad\qquad\qquad h(n)x(n)$$

or

$$\mathbf{y} \equiv X\mathbf{h},$$

where

$$\mathbf{h} \equiv \begin{pmatrix} h(0) \\ \vdots \\ h(n) \end{pmatrix}$$

and

The matrix X is called a data matrix. The order of convolution between two sequences can be changed, as mentioned in Chapter 2.

9.2   Inverse Filters and Linear Equations

9.2.1   Linear Equations

The inverse filter is obtained by taking the inverse matrix of H. Inverse filtering can recover the input sequence $x(n)$ from the output sequence $y(n)$. The inverse filtering is equivalent to solving the linear equations

$$\mathbf{y} = H\mathbf{x}$$

where $\mathbf{x}$ is the input sequence vector, which is the unknown solution vector, H is the impulse response matrix, and $\mathbf{y}$ is the output sequence vector.

However, the inverse matrix of H is not available, since the matrix is not square. The number of equations is larger than that of the unknown variables. The number of equations must be equal to that of the unknown variables if we are to get a unique solution set for a set of simultaneous equations. We have to get a set of approximated solutions to obtain the inverse filter.

9.2.2   Least Square Error Method

Regression methods are used for finding a relationship between random variables. Suppose that Table 9.2.1 is an example of observation data for two random variables $\mathbf{x}$ and $\mathbf{y}$, where we can expect some relationship between the two random variables. Figure 9.2.1 shows the scattered plots for those data and the solid line is the linear regression from $\mathbf{x}$ to $\mathbf{y}$. This solid line is obtained so that the sum of the squared errors of the

Table 9.2.1   A set of random data pairs.

x	y
−6	−4.4462
−5	−2.5462
−4	−2.7462
−3	−1.7462
−2	−0.9462
−1	−0.0538
0	1.2538
1	0.6538
2	0.9538
3	1.7538
4	2.5538
5	2.5538
6	2.7620

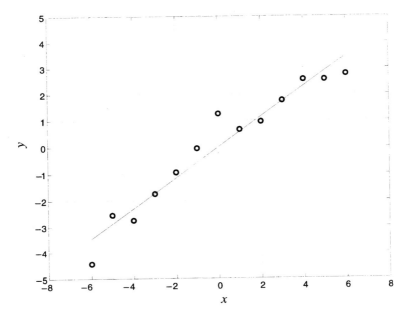

Figure 9.2.1 Random data distribution and a regression line.

observation samples from the solid line becomes minimum. This is called the linear regression method in statistics.

This linear regression analysis provides a method for getting approximated solutions for a set of simultaneous equations. The criterion is to find the least square error (LSE) solution (Lay, 1994). Suppose that we have a set of simultaneous linear equations

$$\mathbf{y} = A\mathbf{x}.$$

The equation above has a set of unique solutions when the matrix A is a square matrix with independent columns. When the matrix A is rectangular (N rows and M columns) and $N < M$, the solutions are available but are not unique. This is because the number of equations N is smaller than that of the unknowns.

If $N > M$, we have no solutions, since the number of equations is larger than that of the unknowns. When $N > M$, the LSE solution $\hat{\mathbf{x}}$ minimizes the squared error which is defined by

$$|\mathbf{e}|^2 \equiv |A\hat{\mathbf{x}} - \mathbf{y}|^2,$$

where $|\ |^2$ denotes the square norm of a vector. The LSE solution $\hat{\mathbf{x}}$ solves the equation

$$\hat{\mathbf{y}} = A\hat{\mathbf{x}},$$

where $\hat{\mathbf{y}} \equiv \mathbf{y} + \mathbf{e}$.

9.2.3   Normal Equation and LSE Solution

The linear equations

$$\mathbf{y} = A\mathbf{x}$$

can be written as a linear combination of the column vectors of the matrix A,

$$\mathbf{y} = x_1\mathbf{a}_1 + x_2\mathbf{a}_2 + \cdots + x_M\mathbf{a}_M,$$

where

$$A \equiv (\,\mathbf{a}_1 \quad \mathbf{a}_2 \quad \ldots \quad \mathbf{a}_M\,),$$

$$\mathbf{x} \equiv \begin{pmatrix} x_1 \\ x_2 \\ \vdots \\ x_M \end{pmatrix} \quad \text{and} \quad \mathbf{y} \equiv \begin{pmatrix} y_1 \\ y_2 \\ \vdots \\ y_N \end{pmatrix}.$$

When the vector $\mathbf{y}$ is a vector in the column space, the combination coefficients vector $\mathbf{x}$ gives the solution vector for the simultaneous equations. When $\mathbf{y}$ is not located in the column space, it cannot be expressed as a linear combination of the column vectors. The orthogonal projection vector $\hat{\mathbf{y}}$ of vector $\mathbf{y}$ onto the column space shown in Fig. 9.2.2a meets the LSE criterion.

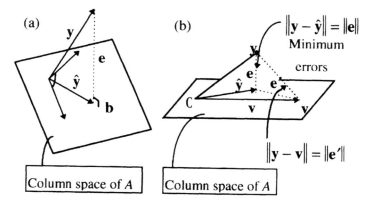

Figure 9.2.2   Orthogonal projection onto the column space: (a) orthogonal projection vector; (b) orthogonal projection vector and least square error criterion.

The squared length of the error vector,

$$|\mathbf{e}|^2 = |\hat{\mathbf{y}} - \mathbf{y}|^2$$

is minimum when $\mathbf{e} \perp \hat{\mathbf{y}}$. Here if we set

$$\hat{\mathbf{y}} = \begin{pmatrix} \hat{y}_1 \\ \hat{y}_2 \\ \vdots \\ \hat{y}_N \end{pmatrix} \equiv \hat{x}_1\mathbf{a}_1 + \hat{x}_2\mathbf{a}_2 + \cdots + \hat{x}_M\mathbf{a}_M \quad \text{and} \quad \hat{\mathbf{x}} \equiv \begin{pmatrix} \hat{x}_1 \\ \hat{x}_2 \\ \vdots \\ \hat{x}_N \end{pmatrix},$$

then $\hat{\mathbf{x}}$ is the least square error (LSE) solution vector which satisfies the linear equations

$$A\hat{\mathbf{x}} = \hat{\mathbf{y}},$$

where

$$\begin{pmatrix} \hat{y}_1 \\ \hat{y}_2 \\ \vdots \\ \hat{y}_N \end{pmatrix} = (\mathbf{a}_1 \quad \mathbf{a}_2 \quad \ldots \quad \mathbf{a}_M) \begin{pmatrix} \hat{x}_1 \\ \hat{x}_2 \\ \vdots \\ \hat{x}_M \end{pmatrix} = A\hat{\mathbf{x}}$$

and

$$A \equiv (\mathbf{a}_1 \quad \mathbf{a}_2 \quad \ldots \quad \mathbf{a}_M).$$

The distance between the vectors $\mathbf{y}$ and $\hat{\mathbf{y}}$ that satisfies the orthogonal relationship $\mathbf{e} \perp \hat{\mathbf{y}}$, is the minimum of the distances between the vector $\mathbf{y}$ and any other vector in the column space, as shown in Fig. 9.2.2b. The orthogonal relationship $\mathbf{e} \perp \hat{\mathbf{y}}$ can be rewritten using the inner product as

$$\hat{\mathbf{y}}^T\mathbf{e} = 0,$$

or equivalently

$$\mathbf{a}_1^T\mathbf{e} = \mathbf{a}_2^T\mathbf{e} = \ldots \mathbf{a}_M^T\mathbf{e} = 0,$$

where

$$\hat{\mathbf{y}} - \mathbf{y} = \mathbf{e},$$

and ^{T} denotes taking the transpose. Substituting the error vector $\mathbf{e}$ into the equations

$$\mathbf{a}_1^T\mathbf{e} = \mathbf{a}_2^T\mathbf{e} = \ldots \mathbf{a}_M^T\mathbf{e} = 0,$$

we can get a set of linear equations given by

$$\begin{pmatrix} \mathbf{a}_1^T \\ \mathbf{a}_2^T \\ \vdots \\ \mathbf{a}_M^T \end{pmatrix} \begin{pmatrix} \hat{y}_1 \\ \hat{y}_2 \\ \vdots \\ \hat{y}_N \end{pmatrix} = \begin{pmatrix} \mathbf{a}_1^T \\ \mathbf{a}_2^T \\ \vdots \\ \mathbf{a}_M^T \end{pmatrix} \begin{pmatrix} y_1 \\ y_2 \\ \vdots \\ y_N \end{pmatrix},$$

where $N > M$. Thus, we can get the normal equations in a matrix form

$$A^TA\hat{\mathbf{x}} = A^T\mathbf{y},$$

where

$$A^T = \begin{pmatrix} \mathbf{a}_1^T \\ \mathbf{a}_2^T \\ \vdots \\ \mathbf{a}_M^T \end{pmatrix}.$$

Here the matrix A^TA is a square and symmetric matrix. When the square matrix A^TA is not singular, we get the LSE solution as

$$\hat{\mathbf{x}} = (A^TA)^{-1}A^T\mathbf{y}.$$

Linear regression, which was illustrated in Fig. 9.2.1, is an example of finding the LSE solution vector. Table 9.2.2 shows the process for finding the linear regression line by solving the normal equation.

Table 9.2.2   A method for linear regression analysis.

x	y
−6	−4.4462
−5	−2.5462
−4	−2.7462
−3	−1.7462
−2	−0.9462
−1	−0.0538
0	1.2538
1	0.6538
2	0.9538
3	1.7538
4	2.5538
5	2.5538
6	2.7620

$$\mathbf{Ax = b}$$

$$\begin{pmatrix} -6 & 1 \\ -5 & 1 \\ -4 & 1 \\ -3 & 1 \\ -2 & 1 \\ -1 & 1 \\ 0 & 1 \\ 1 & 1 \\ 2 & 1 \\ 3 & 1 \\ 4 & 1 \\ 5 & 1 \\ 6 & 1 \end{pmatrix} \begin{pmatrix} a \\ b \end{pmatrix} = \begin{pmatrix} -4.4462 \\ -2.5462 \\ -2.7462 \\ -1.7462 \\ -0.9462 \\ -0.0538 \\ 1.2538 \\ 0.6538 \\ 0.9538 \\ 1.7538 \\ 2.5538 \\ 2.5538 \\ 2.7620 \end{pmatrix}$$

$$\begin{aligned} A\mathbf{x} &= \mathbf{b} \\ A^TA\hat{\mathbf{x}} &= A^T\mathbf{b} \\ \hat{\mathbf{x}} &= (A^TA)^{-1}A^T\mathbf{b} \end{aligned}$$

Table 9.2.2 (*continued*)

$$A^T A = \begin{pmatrix} -6 & -5 & -4 & \cdots & 4 & 5 & 6 \\ 1 & 1 & 1 & \cdots & 1 & 1 & 1 \end{pmatrix} \begin{pmatrix} -6 & 1 \\ -5 & 1 \\ -4 & 1 \\ \vdots & \vdots \\ 4 & 1 \\ 5 & 1 \\ 6 & 1 \end{pmatrix} = \begin{pmatrix} 182 & 0 \\ 0 & 13 \end{pmatrix}$$

$$(A^T A)^{-1} = \frac{1}{2366} \begin{pmatrix} 13 & 0 \\ 0 & 182 \end{pmatrix} = \begin{pmatrix} \dfrac{1}{182} & 0 \\ 0 & \dfrac{1}{13} \end{pmatrix}$$

$$A^T \mathbf{b} = \begin{pmatrix} -6 & -5 & -4 & \cdots & 4 & 5 & 6 \\ 1 & 1 & 1 & \cdots & 1 & 1 & 1 \end{pmatrix} \begin{pmatrix} -4.4462 \\ -2.5462 \\ -2.7462 \\ \vdots \\ 2.5538 \\ 2.5538 \\ 2.7620 \end{pmatrix} = \begin{pmatrix} 104.9568 \\ 0 \end{pmatrix}$$

$$\hat{\mathbf{x}} = (A^T A)^{-1} A^T \mathbf{b} = \begin{pmatrix} \dfrac{1}{182} & 0 \\ 0 & \dfrac{1}{13} \end{pmatrix} \begin{pmatrix} 104.9568 \\ 0 \end{pmatrix} = \begin{pmatrix} 0.5767 \\ 0 \end{pmatrix}$$

$$y = 0.5767x$$

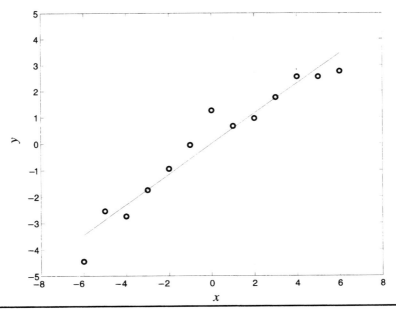

9.2.4   Inverse Filtering Using the LSE Solution

LSE solutions can be applied to get an inverse filter which has a finite impulse response. Suppose that our system impulse response is

$$\mathbf{h} \equiv \begin{pmatrix} h(0) \\ \vdots \\ h(n) \end{pmatrix}.$$

When the input signal sequence vector is $\mathbf{x}$,

$$\mathbf{x} \equiv \begin{pmatrix} x(0) \\ \vdots \\ x(n) \end{pmatrix}$$

the output signal sequence vector $\mathbf{y}$ is given by

$$\mathbf{y} \equiv H\mathbf{x},$$

where

$$\mathbf{y} = \begin{pmatrix} y(0) \\ y(1) \\ \vdots \\ y(n) \\ y(n+1) \\ \vdots \\ y(2n) \end{pmatrix}$$

and

Inverse filtering is recovering the input signal $\mathbf{x}$ from the output signal $\mathbf{y}$ by solving the simultaneous equation

$$\mathbf{y} = H\mathbf{x},$$

where $\mathbf{x}$ is the unknown vector. Since the matrix H is not square, the solution is not available.

Let us take the LSE solution as

$$\hat{\mathbf{y}} = H\hat{\mathbf{x}} \quad \text{and} \quad \hat{\mathbf{x}} = (H^{\mathrm{T}}H)^{-1}H^{\mathrm{T}}\mathbf{y}.$$

This solution gives only the approximate input signal; however, this LSE solution can be obtained even under the condition that the TF, which is the Fourier transform of the impulse response, is not of minimum phase.

9.3 Signal Analysis and Representation

9.3.1 Vector Decomposition of a Discrete Signal

In general, discrete signal analysis can be formulated by a set of simultaneous equations. A discrete signal can be represented as a vector. Suppose that we have a record of N points which are sampled from a continuous signal. The discretized signal is written as a vector in an N-dimensional space. An N-dimensional vector can be expressed by a linear combination of N-dimensional vectors. Decomposition of a signal into the linear combination of a finite number of components is called signal representation.

9.3.2 Orthogonal Vector Decomposition of a Discrete Signal

The most important discrete signal representation is discrete Fourier analysis. Sinusoidal wave components are taken for the vectors which make up a linear combination in order to represent a signal of interest. The discrete Fourier analysis can be formulated as a set of simultaneous equations by

$$A\mathbf{x} = \mathbf{b}.$$

Here

$$A \equiv (\mathbf{a}_1 \quad \mathbf{a}_2 \quad \ldots \quad \mathbf{a}_M)$$

is the matrix of which the columns are M-dimensional vectors composed of M discretized sinusoidal wave components, and $\mathbf{x}$ is the coefficient vector

$$\mathbf{x} \equiv \begin{pmatrix} x_1 \\ x_2 \\ \vdots \\ x_M \end{pmatrix}$$

and

$$\mathbf{b} \equiv \begin{pmatrix} b_1 \\ b_2 \\ \vdots \\ b_M \end{pmatrix}$$

is an M-dimensional signal vector which we want to analyze.

Fourier analysis involves finding the coefficient vector $\mathbf{x}$ which satisfies

$$\mathbf{a}_1 x_1 + \mathbf{a}_2 x_2 + \cdots + \mathbf{a}_M x_M = \mathbf{b}$$

by solving the simultaneous equations

$$A\mathbf{x} = \mathbf{b}.$$

In discrete Fourier analysis, the matrix A is square and all the columns are orthonormal. Thus the simultaneous equations can be solved as

$$\mathbf{x} = A^{-1}\mathbf{b} = A^{\mathrm{T}}\mathbf{b}$$

where $A^{-1} = A^{\mathrm{T}}$ for an orthonormal square matrix A. The coefficient vector can be written as

$$\mathbf{x} = \begin{pmatrix} x_1 \\ x_2 \\ \vdots \\ x_M \end{pmatrix} = (\,\mathbf{a}_1 \quad \mathbf{a}_2 \quad \ldots \quad \mathbf{a}_M\,)^{\mathrm{T}}\mathbf{b} = (\,\mathbf{a}_1 \quad \mathbf{a}_2 \quad \ldots \quad \mathbf{a}_M\,)^{\mathrm{T}} \begin{pmatrix} b_1 \\ b_2 \\ \vdots \\ b_M \end{pmatrix}$$

$$= \begin{pmatrix} \mathbf{a}_1^{\mathrm{T}} \\ \mathbf{a}_2^{\mathrm{T}} \\ \vdots \\ \mathbf{a}_M^{\mathrm{T}} \end{pmatrix} \begin{pmatrix} b_1 \\ b_2 \\ \vdots \\ b_M \end{pmatrix} = \begin{pmatrix} \mathbf{a}_1^{\mathrm{T}}\mathbf{b} \\ \mathbf{a}_2^{\mathrm{T}}\mathbf{b} \\ \vdots \\ \mathbf{a}_M^{\mathrm{T}}\mathbf{b} \end{pmatrix}.$$

Consequently, the signal vector is represented by

$$\mathbf{b} = (\mathbf{a}_1^{\mathrm{T}}\mathbf{b})\mathbf{a}_1 + (\mathbf{a}_2^{\mathrm{T}}\mathbf{b})\mathbf{a}_2 + \cdots + (\mathbf{a}_M^{\mathrm{T}}\mathbf{b})\mathbf{a}_M,$$

where

$$\mathbf{a}_1^{\mathrm{T}}\mathbf{a}_1 = \mathbf{a}_2^{\mathrm{T}}\mathbf{a}_2 = \cdots = \mathbf{a}_M^{\mathrm{T}}\mathbf{a}_M = 1.$$

The signal representation above is called the orthogonal projection onto the column space of the matrix A.

9.3.3 LSE Solution for Orthogonal Vector Decomposition

Discrete Fourier analysis is equivalent to solving the set of simultaneous equations

$$A\mathbf{x} = \mathbf{b},$$

where A is a square matrix composed of orthonormal column vectors and $\mathbf{b}$ is the signal vector to be analyzed. We can generalize the Fourier square orthonormal matrix as a rectangular matrix of which the columns are orthonormal. Suppose that A is a rectangular matrix whose columns are orthonormal. The simultaneous equation is written as

$$A\mathbf{x} = \mathbf{b},$$

where

$$\begin{pmatrix} a_{11} & a_{21} & \cdots & a_{M1} \\ a_{12} & a_{22} & \cdots & a_{M2} \\ \vdots & \vdots & \vdots & \vdots \\ \vdots & \vdots & \vdots & \vdots \\ a_{1N} & a_{2N} & \cdots & a_{MN} \end{pmatrix} \begin{pmatrix} x_1 \\ x_2 \\ \vdots \\ x_M \end{pmatrix} = \begin{pmatrix} b_1 \\ b_2 \\ \vdots \\ \vdots \\ b_N \end{pmatrix}$$

and $N > M$

The matrix A is not square. Only when the signal vector $\mathbf{b}$ is in the column space of A can the equation be solved by

$$\mathbf{x} = \begin{pmatrix} \mathbf{a}_1^T \mathbf{b} \\ \mathbf{a}_2^T \mathbf{b} \\ \vdots \\ \mathbf{a}_M^T \mathbf{b} \end{pmatrix}$$

as well as conventional Fourier analysis. If the signal vector is not located in the column space, the simultaneous equation cannot be solved. However, the orthogonal projection on the column space provides the LSE solution for a signal vector decomposition.

The orthogonal projection $\hat{\mathbf{b}}$ of the signal vector $\mathbf{b}$ onto the orthonormal column space of A is written as

$$\hat{\mathbf{b}} = (\mathbf{a}_1^T \mathbf{b})\mathbf{a}_1 + (\mathbf{a}_2^T \mathbf{b})\mathbf{a}_2 + \cdots + (\mathbf{a}_M^T \mathbf{b})\mathbf{a}_M.$$

The error vector between the signal vector $\mathbf{b}$ outside the column space and the projection vector $\hat{\mathbf{b}}$,

$$\mathbf{e} = \hat{\mathbf{b}} - \mathbf{b},$$

is minimum, since the error vector is orthogonal to the orthonormal column space as shown in Fig. 9.3.1. The most significant vector which approximates the signal following the LSE criterion is the column vector which has the maximum squared inner product with the signal vector.

9.3.4 Speech Signal Decomposition

Speech analysis can be performed on the basis of the discrete Fourier analysis, but most of the speech information is conveyed in the dynamical changes of waveforms with respect to time. Therefore, the discrete Fourier analysis is made on the short time frame basis, say 20–30 ms. It is interesting to see how many sinusoidal components are required to effectively approximate speech signals.

Understandable speech signals can be heard even when only the most significant sinusoidal components are used in each short time frame. Figure 9.3.2b shows a reconstructed speech signal using the most

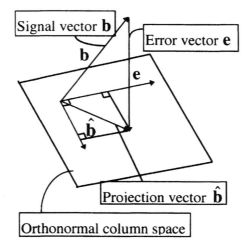

Figure 9.3.1   Orthogonal projection onto the orthonormal column space and error vector.

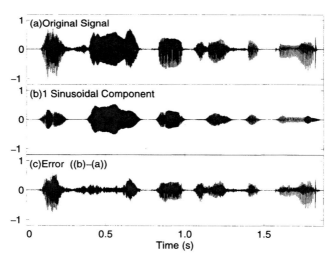

Figure 9.3.2   An example of speech waveform analysis and synthesis: (a) original waveform; (b) synthesized waveform using the most significant sinusoidal component in every short-time frame with 50% overlap; (c) Residual error signal (b) − (a).

significant sinusoidal component in every short time frame with 50% overlap. We can see that the dynamical structures of original signal envelope are reconstructed. Figure 9.3.2c is the error signal between the original and reconstructed waveforms. Figure 9.3.3 demonstrates another example of analysis and synthesis using the five most significant sinusoidal components. We can see that the residual error decreases.

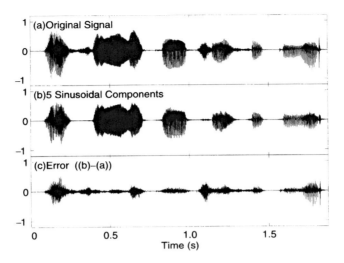

Figure 9.3.3   Analysis and synthesis using the five most significant components in each frame with 50% overlap: (a) original speech waveform; (b) synthesized waveform; (c) residual error.

9.3.5   Generalized Harmonic Analysis

More generalized Fourier analysis can be performed using nonorthogonal column vectors in the simultaneous equation

$$A\mathbf{x} = \mathbf{b},$$

where

$$\begin{pmatrix} a_{11} & a_{21} & \cdots & a_{M1} \\ a_{12} & a_{22} & \cdots & a_{M2} \\ \vdots & \vdots & \vdots & \vdots \\ \vdots & \vdots & \vdots & \vdots \\ a_{1N} & a_{2N} & \cdots & a_{MN} \end{pmatrix} \begin{pmatrix} x_1 \\ x_2 \\ \vdots \\ x_M \end{pmatrix} = \begin{pmatrix} b_1 \\ b_2 \\ \vdots \\ \vdots \\ b_N \end{pmatrix}.$$

We can take the LSE solution

$$\hat{\mathbf{x}} = \begin{pmatrix} \hat{x}_1 \\ \hat{x}_2 \\ \vdots \\ \hat{x}_M \end{pmatrix} = (A^{\mathrm{T}}A)^{-1}A^{\mathrm{T}}\mathbf{b}$$

for signal representation, where $A\hat{\mathbf{x}} \equiv \hat{\mathbf{b}}$ and $\hat{\mathbf{b}}$ denotes the orthogonal projection vector of the signal vector $\mathbf{b}$ onto the nonorthogonal column space of A. This is called generalized harmonic analysis (GHA). In terms of acoustics, this is signal representation using nonharmonic frequency components, while conventional Fourier analysis always uses a set of harmonic frequency components (Terada *et al*., 1994).

Suppose we have an N-dimensional vector whose entries are sampled from a sinusoidal wave of single frequency. The discrete Fourier transform (DFT) transforms the N sampled data to N discretized frequency components. The DFT represents a single sinusoidal wave as the summation of N frequency components. Only when the number of data points N of the sinusoidal wave is an integer multiple of the period can the sinusoidal wave be expressed by the single frequency component. However, this condition is rare in usual discrete Fourier analysis.

We can select the column vectors flexibly in the generalized harmonic analysis in order to represent a signal using only a few vectors. If the signal vector $\mathbf{b}$ is located in the column space, then the error becomes zero and the signal can be expressed as

$$\hat{\mathbf{b}} = \mathbf{b} = \hat{x}_1\mathbf{a}_1 + \hat{x}_2\mathbf{a}_2 + \cdots + \hat{x}_M\mathbf{a}_M = A\hat{\mathbf{x}} = A\mathbf{x}.$$

However, the signal is not expressed as, even if the signal vector is located in the column space,

$$A\hat{\mathbf{x}} = \hat{\mathbf{b}} = \mathbf{b} = (\mathbf{a}_1^{\mathrm{T}}\mathbf{b})\mathbf{a}_1 + (\mathbf{a}_2^{\mathrm{T}}\mathbf{b})\mathbf{a}_2 + \cdots + (\mathbf{a}_M^{\mathrm{T}}\mathbf{b})\mathbf{a}_M,$$

since the column vectors are not orthogonal. Figure 9.3.4 shows the geometric interpretation of orthogonal projection onto the nonorthogonal column space. We take the solution as

$$\hat{\mathbf{x}} = (A^{\mathrm{T}}A)^{-1}A^{\mathrm{T}}\mathbf{b} = A^{-1}\mathbf{b}$$

for GHA in the nonorthogonal column space.

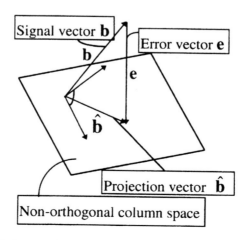

Figure 9.3.4 Orthogonal projection onto the nonorthogonal column space.

9.4   Principal Component Analysis

9.4.1   Orthogonal Regression Method

We described the least square error method in relation to linear regression in Section 9.2. In statistics, we have another regression method called orthogonal regression. This regression is also related to a form of signal analysis called principal component analysis (PCA). Suppose that Table 9.4.1 is an example of observation data of two random variables **x** and **y**. Figure 9.4.1 shows the scatter plots for those data and the solid line is obtained so that the sum of the squared orthogonal distances between the observation samples and the solid line becomes minimum. This is called orthogonal regression analysis in statistics.

The orthogonal regression method uses the orthogonal projection. Figure 9.4.2 shows the geometric interpretation of such a projection. Suppose that we have N sampled data points which are plotted in the 2-dimensional xy-plane. The ith sample is located at (x_i, y_i) and we define the ith sample vector as $\mathbf{v}_i$. Both x_i and y_i are random variables of zero mean. The solid line in the figure is a candidate for the orthogonal regression line.

We have the relationship

$$\sum_{i=1}^{N} r_i^2 = \sum_{i=1}^{N} (x_i^2 + y_i^2) \equiv \sum_{i=1}^{N} (l_i^2 + h_i^2) = \sum_{i=1}^{N} l_i^2 + \sum_{i=1}^{N} h_i^2 \equiv K,$$

where r_i denotes the length of the ith sample vector, l_i is the projection length of the sample vector onto the solid line,

$$l_i = x_i u_x + y_i u_y = \mathbf{v}_i^{\mathrm{T}} \mathbf{u},$$

Table 9.4.1   A set of random data pairs.

x	y
−6	−4.4462
−5	−2.5462
−4	−2.7462
−3	−1.7462
−2	−0.9462
−1	−0.0538
0	1.2538
1	0.6538
2	0.9538
3	1.7538
4	2.5538
5	2.5538
6	2.7620

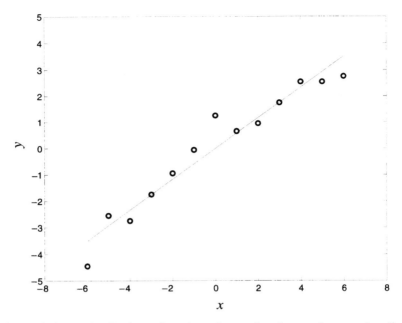

Figure 9.4.1   Distribution of random data and orthogonal regression line.

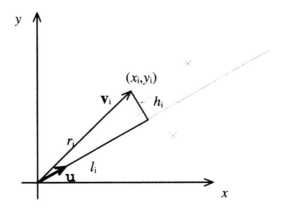

Figure 9.4.2   Geometric interpretation of orthogonal regression.

$\mathbf{u}$ is the unit vector for the direction of the solid line, and h_i is the orthogonal distance between the sample vector and the solid line. Making the sum of the squared orthogonal distance, $\Sigma_{i=1}^{N} h_i^2$, minimum is equivalent to making the sum of the squared projection length, $\Sigma_{i=1}^{N} l_i^2$ maximum, when K is given.

Let us take the data matrix which is composed of N pairs of (x_i, y_i) as

$$B \equiv \begin{pmatrix} x_1 & x_2 & \cdots & x_N \\ y_1 & y_2 & \cdots & y_N \end{pmatrix} \equiv (\mathbf{v}_1 \quad \mathbf{v}_2 \quad \cdots \quad \mathbf{v}_N).$$

The projected vector $\mathbf{l}$ onto a line whose unit vector is $\mathbf{u}$ is

$$\mathbf{l} = B^{\mathrm{T}} \mathbf{u} = \begin{pmatrix} x_1 & x_2 & \cdots & x_N \\ y_1 & y_2 & \cdots & y_N \end{pmatrix}^{\mathrm{T}} \mathbf{u}$$

$$= (\mathbf{v}_1 \quad \mathbf{v}_2 \quad \cdots \quad \mathbf{v}_N)^{\mathrm{T}} \mathbf{u} = \begin{pmatrix} \mathbf{v}_1^{\mathrm{T}} \mathbf{u} \\ \mathbf{v}_2^{\mathrm{T}} \mathbf{u} \\ \vdots \\ \mathbf{v}_N^{\mathrm{T}} \mathbf{u} \end{pmatrix}.$$

The unit vector $\mathbf{u}$ for the orthogonal regression line is determined so that the length of the vector $\mathbf{l}$ is maximized. Such a length maximization can be made based on the characteristics of the quadratic forms of matrices.

9.4.2 Quadratic Forms of Matrices

The matrix is an operator for linear transformations. Matrix A transforms a vector $\mathbf{x}$ in the row space into the column space vector $\mathbf{b}$ as

$$A\mathbf{x} = \mathbf{b}.$$

Suppose that A is a symmetric matrix. The square length of the vector $\mathbf{b}$,

$$\mathbf{b}^{\mathrm{T}}\mathbf{b} = (A\mathbf{x})^{\mathrm{T}}(A\mathbf{x}) = \mathbf{x}^{\mathrm{T}}(A^{\mathrm{T}}A)\mathbf{x},$$

is called the quadratic form of the matrix A, where $(A\mathbf{x})^{\mathrm{T}} = \mathbf{x}^{\mathrm{T}}A^{\mathrm{T}}$.

Let us find a unit vector $\mathbf{u}$ in the row space which makes the quadratic form $\mathbf{u}^{\mathrm{T}}(A^{\mathrm{T}}A)\mathbf{u}$ maximum. The quadratic theory says (Lay, 1994) that $\mathbf{u}^{\mathrm{T}}(A^{\mathrm{T}}A)\mathbf{u}$ becomes maximum when we take the unit eigenvector $\mathbf{u}$ which belongs to the maximum eigenvalue of the matrix $A^{\mathrm{T}}A$. If we take a unit eigenvector $\mathbf{u}$ which belongs to the eigenvalue σ^2 of the matrix $A^{\mathrm{T}}A$, the square length of the transformed vector

$$\mathbf{v} \equiv A\mathbf{u}$$

is written as

$$\mathbf{v}^{\mathrm{T}}\mathbf{v} = (A\mathbf{u})^{\mathrm{T}}(A\mathbf{u}) = \mathbf{u}^{\mathrm{T}}(A^{\mathrm{T}}A)\mathbf{u} = \mathbf{u}^{\mathrm{T}}\sigma^2\mathbf{u} = \sigma^2\mathbf{u}^{\mathrm{T}}\mathbf{u} = \sigma^2.$$

Thus, when σ^2 is the maximum eigenvalue of the matrix $A^{\mathrm{T}}A$ and $\mathbf{u}$ is the eigenvector corresponding to this eigenvalue, squared length of the transformed vector **v** becomes maximum. The matrix A^TA is symmetric, and its eigenvalues must be nonnegative, since an eigenvalue of matrix A^TA shows squared length of the transformed vector of the eigenvector for the eigenvalue. The eigenvectors for a symmetric matrix belonging to the different eigenvalues are orthogonal. This is because if we take two eigenvectors $\mathbf{u}_1$ and $\mathbf{u}_2$ of a symmetric matrix A that correspond to distinct eigenvalues λ_1 and λ_2, then we have

$$\lambda_1 \mathbf{u}_1 \cdot \mathbf{u}_2 = (\lambda_1 \mathbf{u}_1)^T \mathbf{u}_2 = (A\mathbf{u}_1)^T \mathbf{u}_2 = (\mathbf{u}_1^T A^T)\mathbf{u}_2 = \mathbf{u}_1^T (A\mathbf{u}_2)$$

$$= \mathbf{u}_1^T (\lambda_2 \mathbf{u}_2) = \lambda_2 \mathbf{u}_1^T \mathbf{u}_2 = \lambda_2 \mathbf{u}_1 \cdot \mathbf{u}_2$$

where $A^T = A$. Therefore, the relation

$$(\lambda_1 - \lambda_2)\mathbf{u}_1 \cdot \mathbf{u}_2 = 0$$

holds and we get $\mathbf{u}_1 \cdot \mathbf{u}_2 = 0$ for $\lambda_1 \neq \lambda_2$. The square roots of the nonnegative eigenvalues of the matrix A^TA are called singular values of the matrix A.

9.4.3   Principal Axis of a Quadratic Form

Figure 9.4.3 shows an example of the transforms $A\mathbf{u} = \mathbf{v}$ from the row space to the column space. The matrix A has a three-dimensional row space given by

$$A = \begin{pmatrix} 4 & 11 & 14 \\ 8 & 7 & -2 \end{pmatrix}.$$

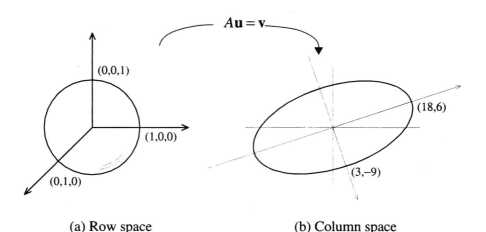

(a) Row space   (b) Column space

Figure 9.4.3   An example of linear transformation from a three-dimensional row space to a two-dimensional column space: (a) row space; (b) column space. (Lay, 1994.)

The unit vectors in the three-dimensional row space make a sphere with unit radius as shown in Fig. 9.4.3a. The matrix A transforms the unit sphere into an ellipse as shown in Fig. 9.4.3b.

The two orthogonal principal axes for the ellipse are transformed from the unit eigenvectors of the matrix A^TA. The matrix A^TA becomes

$$A^TA = \begin{pmatrix} 4 & 8 \\ 11 & 7 \\ 14 & -2 \end{pmatrix} \begin{pmatrix} 4 & 11 & 14 \\ 8 & 7 & -2 \end{pmatrix} = \begin{pmatrix} 80 & 100 & 40 \\ 100 & 170 & 140 \\ 40 & 140 & 200 \end{pmatrix}.$$

The eigenvalues of A^TA are $\sigma_1^2 = 360$, $\sigma_2^2 = 90$, and $\sigma_3^2 = 0$. The corresponding eigenvectors become

$$\mathbf{u}_1 = \begin{pmatrix} \frac{1}{3} \\ \frac{2}{3} \\ \frac{2}{3} \end{pmatrix} \qquad \text{for } \sigma_1^2 = 360,$$

$$\mathbf{u}_2 = \begin{pmatrix} -\frac{2}{3} \\ -\frac{1}{3} \\ \frac{2}{3} \end{pmatrix} \qquad \text{for } \sigma_2^2 = 90,$$

and

$$\mathbf{u}_3 = \begin{pmatrix} \frac{2}{3} \\ -\frac{2}{3} \\ \frac{1}{3} \end{pmatrix} \qquad \text{for } \sigma_3^2 = 0.$$

If we take the eigenvector $\mathbf{u}_1$ which corresponds to the maximum eigenvalue $\sigma_1^2 = 360$ as a three-dimensional unit vector in the row space, we get the transformed vector

$$\mathbf{v}_1 = A\mathbf{u}_1 = \begin{pmatrix} 4 & 11 & 14 \\ 8 & 7 & -2 \end{pmatrix} \begin{pmatrix} \frac{1}{3} \\ \frac{2}{3} \\ \frac{2}{3} \end{pmatrix} = \begin{pmatrix} 18 \\ 6 \end{pmatrix}.$$

The squared length of the vector $\mathbf{v}_1$ is $18^2 + 6^2 = 360$ which is equal to σ_1^2. For the second vector $\mathbf{u}_2$, we get

$$\mathbf{v}_2 = A\mathbf{u}_2 = \begin{pmatrix} 4 & 11 & 14 \\ 8 & 7 & -2 \end{pmatrix} \begin{pmatrix} -\frac{2}{3} \\ -\frac{1}{3} \\ \frac{2}{3} \end{pmatrix} = \begin{pmatrix} 3 \\ -9 \end{pmatrix}.$$

The squared length of the second vector $\mathbf{v}_2$ is 90 equal to σ_2^2. The matrix A transforms the third vector to zero as

$$\mathbf{v}_3 = A\mathbf{u}_3 = \begin{pmatrix} 4 & 11 & 14 \\ 8 & 7 & -2 \end{pmatrix} \begin{pmatrix} \frac{2}{3} \\ -\frac{2}{3} \\ \frac{1}{3} \end{pmatrix} = \begin{pmatrix} 0 \\ 0 \end{pmatrix}.$$

Figure 9.4.3b shows that these transformed vectors are the principal axes and also shows that the vector transformed from the eigenvector which has the maximum eigenvalue is the first principal axis.

9.4.4   Diagonalization of Covariance Matrix

Principal component analysis or orthogonal regression in statistics means finding the principal axes for a random data matrix. We will find the unit vector which maximizes the length of the vector

$$\mathbf{l} \equiv B^{\mathrm{T}}\mathbf{u} \equiv \begin{pmatrix} x_1 & x_2 & \cdots & x_N \\ y_1 & y_2 & \cdots & y_N \end{pmatrix}^{\mathrm{T}} \mathbf{u} \equiv (\,\mathbf{b}_1 \quad \mathbf{b}_2 \quad \cdots \quad \mathbf{b}_N\,)^{\mathrm{T}} \mathbf{u}.$$

The eigenvector corresponding to the largest eigenvalue of matrix

$$(B^{\mathrm{T}})^{\mathrm{T}} B^{\mathrm{T}} = BB^{\mathrm{T}},$$

makes the squared length of the vector $\mathbf{l}$ maximum. Here the matrix BB^{T} is called the covariance matrix of the data matrix B.

The data column vector $\mathbf{b}_i$ can be expressed using the eigenvectors of the covariance matrix, $\mathbf{u}_1$ and $\mathbf{u}_2$, which are orthogonal:

$$\mathbf{b}_i \equiv \alpha_{i1}\mathbf{u}_1 + \alpha_{i2}\mathbf{u}_2 = (\,\mathbf{u}_1 \quad \mathbf{u}_2\,) \begin{pmatrix} \alpha_{i1} \\ \alpha_{i2} \end{pmatrix}$$

where

$$\alpha_{i1} = \mathbf{b}_i^{\mathrm{T}}\mathbf{u}_1 = \mathbf{u}_1^{\mathrm{T}}\mathbf{b}_i \qquad \text{and} \qquad \alpha_{i2} = \mathbf{b}_i^{\mathrm{T}}\mathbf{u}_2 = \mathbf{u}_2^{\mathrm{T}}\mathbf{b}_i.$$

Those eigenvectors $\mathbf{u}_1$ and $\mathbf{u}_2$ are orthogonal, since the covariance matrix BB^{T} is symmetric. The data matrix B

$$B \equiv \begin{pmatrix} x_1 & x_2 & \cdots & x_N \\ y_1 & y_2 & \cdots & y_N \end{pmatrix} \equiv (\,\mathbf{b}_1 \quad \mathbf{b}_2 \quad \cdots \quad \mathbf{b}_N\,),$$

can be rewritten using the eigenvectors of the covariance matrix as

$$B \equiv (\,\mathbf{b}_1 \quad \mathbf{b}_2 \quad \cdots \quad \mathbf{b}_N\,) = (\,\mathbf{u}_1 \quad \mathbf{u}_2\,) \begin{pmatrix} \alpha_{11} & \alpha_{21} & \cdots & \alpha_{N1} \\ \alpha_{12} & \alpha_{22} & \cdots & \alpha_{N2} \end{pmatrix},$$

where

$$\begin{pmatrix} \alpha_{11} & \alpha_{21} & \cdots & \alpha_{N1} \\ \alpha_{12} & \alpha_{22} & \cdots & \alpha_{N2} \end{pmatrix} = (\mathbf{u}_1 \quad \mathbf{u}_2)^{\mathrm{T}} (\mathbf{b}_1 \quad \mathbf{b}_2 \quad \cdots \quad \mathbf{b}_N)$$

$$= (\mathbf{u}_1 \quad \mathbf{u}_2)^{\mathrm{T}} \begin{pmatrix} x_1 & x_2 & \cdots & x_N \\ y_1 & y_2 & \cdots & y_N \end{pmatrix}.$$

These changing variables make the covariance matrix diagonal. The covariance matrix S for the new variables is given by

$$S \equiv \begin{pmatrix} \alpha_{11} & \alpha_{21} & \cdots & \alpha_{N1} \\ \alpha_{12} & \alpha_{22} & \cdots & \alpha_{N2} \end{pmatrix} \begin{pmatrix} \alpha_{11} & \alpha_{21} & \cdots & \alpha_{N1} \\ \alpha_{12} & \alpha_{22} & \cdots & \alpha_{N2} \end{pmatrix}^{\mathrm{T}}$$

$$= (\mathbf{u}_1 \quad \mathbf{u}_2)^{\mathrm{T}} BB^{\mathrm{T}} (\mathbf{u}_1 \quad \mathbf{u}_2),$$

where we can diagonalize the matrix BB^{T} as

$$BB^{\mathrm{T}} = (\mathbf{u}_1 \quad \mathbf{u}_2) D (\mathbf{u}_1 \quad \mathbf{u}_2)^{-1} = (\mathbf{u}_1 \quad \mathbf{u}_2) D (\mathbf{u}_1 \quad \mathbf{u}_2)^{\mathrm{T}},$$

$$(\mathbf{u}_1 \quad \mathbf{u}_2)^{\mathrm{T}} = (\mathbf{u}_1 \quad \mathbf{u}_2)^{-1}$$

and

$$D = \begin{pmatrix} \sigma_1^2 & 0 \\ 0 & \sigma_2^2 \end{pmatrix}.$$

(See Section 9.5.1.) σ_1^2 and σ_2^2 are eigenvalues of the covariance matrix BB^{T}. Therefore, the new covariance matrix S becomes diagonal as

$$S = (\mathbf{u}_1 \quad \mathbf{u}_2)^{\mathrm{T}} BB^{\mathrm{T}} (\mathbf{u}_1 \quad \mathbf{u}_2)$$

$$= (\mathbf{u}_1 \quad \mathbf{u}_2)^{\mathrm{T}} (\mathbf{u}_1 \quad \mathbf{u}_2) D (\mathbf{u}_1 \quad \mathbf{u}_2)^{-1} (\mathbf{u}_1 \quad \mathbf{u}_2) = D.$$

The diagonal entries, which are the eigenvalues of the covariance matrix, are the variances of the random variables. The eigenvalues of a covariance matrix are nonnegative. If the variances of random variables are concentrated in significant diagonal entries, the set of multidimensional random data vectors can be characterized by the significant eigenvectors. We can reduce the dimension of the random vector space.

When the covariance matrix is diagonal, there are no correlations between any pair of random variables which are entries of the column vectors of the random data matrix. The newly transformed random variables are uncorrelated to each other; therefore, we can diagonalize the covariance matrix for the new random variables.

Table 9.4.2 shows the process for getting the orthogonal regression line shown in Fig. 9.4.1 for the data matrix B. The data plot originally looks

Table 9.4.2   The process for orthogonal regression analysis.

Data (with zero average) matrix

$$B = \begin{bmatrix} -6 & -5 & \cdots & 5 & 6 \\ -4.4462 & -1.5462 & \cdots & 2.5538 & 2.7620 \end{bmatrix}$$

Covariance matrix

$$S = \frac{1}{N-1} BB^{\mathrm{T}}$$

$$= \frac{1}{13-1} \begin{bmatrix} -6 & -5 & \cdots & 5 & 6 \\ -4.4462 & -1.5462 & \cdots & 2.5538 & 2.7620 \end{bmatrix} \begin{bmatrix} -6 & -4.4462 \\ -5 & -1.5462 \\ \vdots & \vdots \\ 5 & 2.5538 \\ 6 & 2.7620 \end{bmatrix}$$

$$= \begin{bmatrix} 15.1666 & 8.7464 \\ 8.7464 & 5.3665 \end{bmatrix}$$

Eigenvalues of S

$$\lambda_1 = 20.291, \quad \lambda_2 = 0.2411$$

The first principal component vector

$$\mathbf{v}_1 = \begin{bmatrix} -0.8628 \\ -0.5056 \end{bmatrix} \ldots\ldots \text{First principal component (normalized)}$$

---

as if it is 2-dimensionally distributed; however, the data can be seen to be almost one-dimensionally distributed following the first principal axis. The components for the vertical axis of the data plots are orthogonal projections from the column vectors of the data matrix onto the second eigenvector.

9.4.5   Principal Component Analysis of Random Data

The covariance matrix characterizes the random nature of signals. Random signals are unpredictable in a deterministic way. The spectrum is a random variable for a short-time window. Let us take the discrete Fourier transforms of a series of short-time windows and describe the data matrix for the magnitude spectra by

$$B \equiv \begin{pmatrix} p_{11} & \cdots & p_{M1} \\ \vdots & \vdots & \vdots \\ \vdots & \vdots & \vdots \\ p_{1N} & \cdots & p_{MN} \end{pmatrix},$$

where N denotes the number of frequency components and M the number of short time frames. The column vector of the matrix shows a random vector in an N-dimensional space. The magnitude spectrum is a random quantity which fluctuates as the time frame changes. Such fluctuations are observed in the entries of a row vector in matrix B. The random signal signature can be characterized by the spectrum fluctuations and their correlation among the frequency components (row vectors).

The orthogonal regression analysis can be extended into multidimensional space just as the linear regression is extended into the LSE method. The principal vector for representing the major direction of the random vector in the multidimensional distribution can be found. This is a key technology for dimension analysis of random data. The random variables given by an N-dimensional vector require N independent vectors to be represented. If we can find several candidates for the significant vectors which represent any N-dimensional random vectors of the signal of interest, then we can reduce the dimension of the space constructed by the random vector.

Figure 9.4.4 demonstrates the covariance matrix for a short-time spectrum. Figure 9.4.4a shows the random noise spectra and Fig. 9.4.4b illustrates the speech spectra. The matrix for the random noise is almost diagonal, since there is little correlation in the frequency components of the spectra. On the other hand, the covariance matrix for the speech signal has nonzero entries also for nondiagonal terms.

The variances of random variables are represented by the eigenvalues of the covariance matrix. Figure 9.4.5 shows the eigenvalue distribution of the covariance matrix where the eigenvalues are arranged in order from maximum to minimum magnitude. The magnitudes of the eigenvalues for the noise spectrum are rather uniformly distributed (dotted line), while those for speech are concentrated in several significant eigenvalues (solid line). A speech signal can be reconstructed approximately using only those principal vectors which have significant eigenvalues for the spectrum frame variation.

If we take the orthogonal projections of the column vectors $\mathbf{p}$ onto the principal eigenvectors $\mathbf{u}$, then we can express the ith power spectrum column vector $\mathbf{p}_i$ as

$$\mathbf{p}_i = \beta_{i1}\mathbf{u}_1 + \beta_{i2}\mathbf{u}_2 + \cdots + \beta_{iN}\mathbf{u}_N,$$

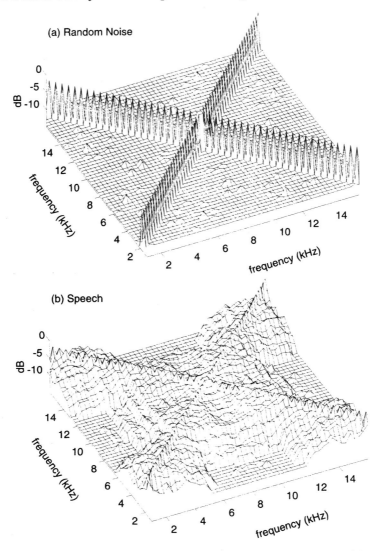

Figure 9.4.4   Examples of covariance matrix for power spectrum: (a) random noise; (b) speech signal.

where

$$B \equiv \begin{pmatrix} p_{11} & \cdots & p_{M1} \\ \vdots & \vdots & \vdots \\ \vdots & \vdots & \vdots \\ p_{1N} & \cdots & p_{MN} \end{pmatrix} \equiv (\mathbf{p}_1 \quad \cdots \quad \mathbf{p}_M).$$

Figure 9.4.6 expresses the reconstructed speech waveform using 20 principal vectors for each 512-dimensional spectrum vector with the

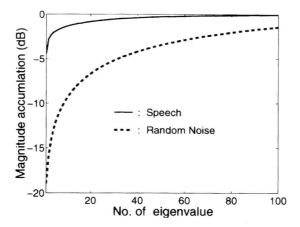

Figure 9.4.5 Normalized accumulation of eigenvalue magnitudes for the covariance matrix: broken line, (a) broken random noise; (b) solid line speech signal.

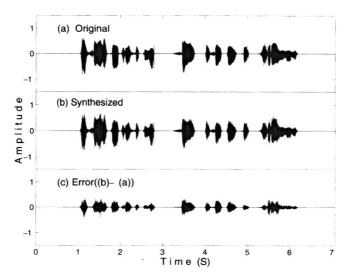

Figure 9.4.6 Reconstructed speech waveform using the significant eigenvectors.

original phase spectrum. Almost all dynamical features of the speech entire waveform can be recovered.

9.5 Singular Value Decomposition of a Matrix

9.5.1 Orthogonal Diagonalization of a Symmetric Matrix

The nonnegative square roots of the eigenvalues of the matrix $A^T A$ are called singular values of the matrix A. A matrix can be diagonalized using the singular values. This is called singular value decomposition. When a matrix A has the relationship

$$A\mathbf{x} = \lambda\mathbf{x}$$

with a vector, the vector $\mathbf{x}$ is called an eigenvector of the matrix A and λ is an eigenvalue of the matrix A. Figure 9.5.1 shows a geometric interpretation of the linear transformation in a two-dimensional space.

$$\begin{pmatrix} 3 & -2 \\ 1 & 0 \end{pmatrix}\begin{pmatrix} -1 \\ 1 \end{pmatrix} = (-1)\begin{pmatrix} 3 \\ 1 \end{pmatrix} + (1)\begin{pmatrix} -2 \\ 0 \end{pmatrix} = \begin{pmatrix} -5 \\ -1 \end{pmatrix}$$

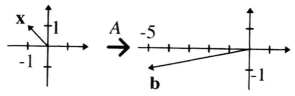

Figure 9.5.1   Linear transformation of a vector by a matrix.

A square matrix A can be written with the eigenvectors $\mathbf{x}_i$, or equivalently

$$AX = (\mathbf{x}_1 \quad \cdots \quad \mathbf{x}_n)\begin{pmatrix} \lambda_1 & \cdots & 0 \\ \vdots & \ddots & \vdots \\ 0 & \cdots & \lambda_n \end{pmatrix} = X\Lambda,$$

where

$$\begin{pmatrix} \lambda_1 & \cdots & 0 \\ \vdots & \ddots & \vdots \\ 0 & \cdots & \lambda_n \end{pmatrix} \equiv \Lambda.$$

When the matrix A is symmetric, the eigenvectors are orthogonal. Thus, using the relationship above, the symmetric matrix A is orthogonally diagonalized as

$$AXX^{-1} = A = X\Lambda X^{-1}$$

$$= (\mathbf{x}_1 \quad \cdots \quad \mathbf{x}_n)\begin{pmatrix} \lambda_1 & \cdots & 0 \\ \vdots & \ddots & \vdots \\ 0 & \cdots & \lambda_n \end{pmatrix}(\mathbf{x}_1 \quad \cdots \quad \mathbf{x}_n)^{-1}.$$

Then we can write the inverse matrix as

$$A^{-1} = (X \Lambda X^{-1})^{-1} = X \Lambda^{-1} X^{-1}$$

$$= (\mathbf{x}_1 \quad \dots \quad \mathbf{x}_n) \begin{pmatrix} \lambda_1^{-1} & \dots & 0 \\ \vdots & \ddots & \vdots \\ 0 & \dots & \lambda_n^{-1} \end{pmatrix} \begin{pmatrix} \mathbf{x}_1^{\mathrm{T}} \\ \vdots \\ \mathbf{x}_n^{\mathrm{T}} \end{pmatrix},$$

where

$$\Lambda^{-1} \equiv \begin{pmatrix} \lambda_1 & \dots & 0 \\ \vdots & \ddots & \vdots \\ 0 & \dots & \lambda_n \end{pmatrix}^{-1} = \begin{pmatrix} \lambda_1^{-1} & \dots & 0 \\ \vdots & \ddots & \vdots \\ 0 & \dots & \lambda_n^{-1} \end{pmatrix}$$

and

$$X^{-1} \equiv (\mathbf{x}_1 \quad \dots \quad \mathbf{x}_n)^{-1} = (\mathbf{x}_1 \quad \dots \quad \mathbf{x}_n)^{\mathrm{T}},$$

since the eigenvectors are orthogonal for the symmetric matrix. Therefore, the inverse transform is written as

$$A^{-1}AX = (\mathbf{x}_1 \quad \dots \quad \mathbf{x}_n) \begin{pmatrix} \lambda_1^{-1} & \dots & 0 \\ \vdots & \ddots & \vdots \\ 0 & \dots & \lambda_n^{-1} \end{pmatrix} \begin{pmatrix} \mathbf{x}_1^{\mathrm{T}} \\ \vdots \\ \mathbf{x}_n^{\mathrm{T}} \end{pmatrix}$$

$$\times (\mathbf{x}_1 \quad \dots \quad \mathbf{x}_n) \begin{pmatrix} \lambda_1 & \dots & 0 \\ \vdots & \ddots & \vdots \\ 0 & \dots & \lambda_n \end{pmatrix}$$

$$= X.$$

9.5.2 Diagonalization and Invertibility of a Matrix $A^{\mathrm{T}}A$

When the matrix A is rectangular, the inverse transform does not hold, since the matrix does not have the eigenvalues and eigenvectors. A matrix, however, can be diagonalized using its singular values. Let us take a rectangular matrix A,

$$A \equiv \begin{pmatrix} a_{11} & a_{21} & \dots & a_{M1} \\ a_{12} & a_{22} & \dots & a_{M2} \\ \vdots & \vdots & \dots & \vdots \\ \vdots & \vdots & \dots & \vdots \\ a_{1N} & a_{2N} & \dots & a_{MN} \end{pmatrix}.$$

The matrix $A^{\mathrm{T}}A$,

$$A^{\mathrm{T}}A = \begin{pmatrix} \mathbf{a}_1^{\mathrm{T}} \\ \mathbf{a}_2^{\mathrm{T}} \\ \vdots \\ \mathbf{a}_M^{\mathrm{T}} \end{pmatrix} (\mathbf{a}_1 \quad \mathbf{a}_2 \quad \dots \quad \mathbf{a}_M) = \begin{pmatrix} \mathbf{a}_1^{\mathrm{T}}\mathbf{a}_1 & \mathbf{a}_1^{\mathrm{T}}\mathbf{a}_2 & \dots & \mathbf{a}_1^{\mathrm{T}}\mathbf{a}_M \\ \mathbf{a}_2^{\mathrm{T}}\mathbf{a}_1 & \mathbf{a}_2^{\mathrm{T}}\mathbf{a}_2 & \dots & \mathbf{a}_2^{\mathrm{T}}\mathbf{a}_M \\ \vdots & \vdots & \vdots & \vdots \\ \mathbf{a}_M^{\mathrm{T}}\mathbf{a}_1 & \mathbf{a}_M^{\mathrm{T}}\mathbf{a}_2 & \dots & \mathbf{a}_M^{\mathrm{T}}\mathbf{a}_M \end{pmatrix}$$

is symmetric and can be orthogonally diagonalized using the orthogonal eigenvectors

$$A^{\mathrm{T}}A = U\Sigma U^{-1} = U\Sigma U^{\mathrm{T}}$$

$$= (\mathbf{u}_1 \quad \dots \quad \mathbf{u}_M) \begin{pmatrix} \sigma_1^2 & \dots & 0 \\ \vdots & \ddots & \vdots \\ 0 & \dots & \sigma_M^2 \end{pmatrix} (\mathbf{u}_1 \quad \dots \quad \mathbf{u}_M)^{\mathrm{T}},$$

where $\mathbf{u}_i$ denotes the ith eigenvector of the matrix $A^{\mathrm{T}}A$ and $\Sigma \equiv \begin{pmatrix} \sigma_1^2 & \dots & 0 \\ \vdots & \ddots & \vdots \\ 0 & \dots & \sigma_M^2 \end{pmatrix}$. Therefore, as already discussed when we used the inverse matrix to get the LSE solutions, the inverse of the matrix $A^{\mathrm{T}}A$ can be written as

$$(A^{\mathrm{T}}A)^{-1} = (U\Sigma U^{-1})^{-1} = U\Sigma^{-1}U^{-1} = U\Sigma^{-1}U^{\mathrm{T}}$$

$$= (\mathbf{u}_1 \quad \dots \quad \mathbf{u}_M) \begin{pmatrix} \sigma_1^{-2} & \dots & 0 \\ \vdots & \ddots & \vdots \\ 0 & \dots & \sigma_M^{-2} \end{pmatrix} (\mathbf{u}_1 \quad \dots \quad \mathbf{u}_M)^{\mathrm{T}}.$$

The inverse matrix of $A^{\mathrm{T}}A$ is not always available, since the eigenvalues can be zeros.

A square matrix B with eigenvector $\mathbf{x}$ and eigenvalue λ satisfies

$$B\mathbf{x} = \lambda\mathbf{x}.$$

If the matrix B has a zero eigenvalue, then we have a solution for the equation

$$B\mathbf{x} = 0.$$

This equation is called a homogeneous linear equation. If we have the nontrivial (nonzero vector) solution $\mathbf{x}$ for this homogeneous equation,

$$\begin{pmatrix} b_{11} & b_{21} & \dots & b_{N1} \\ b_{12} & b_{22} & \dots & b_{N2} \\ \vdots & \vdots & \vdots & \vdots \\ b_{1N} & b_{2N} & \dots & b_{NN} \end{pmatrix} \begin{pmatrix} x_1 \\ x_2 \\ \vdots \\ x_N \end{pmatrix} \equiv (\mathbf{b}_1 \quad \mathbf{b}_2 \quad \dots \quad \mathbf{b}_N) \begin{pmatrix} x_1 \\ x_2 \\ \vdots \\ x_N \end{pmatrix} = 0,$$

then the column vectors are not linearly independent, such that

$$x_1 \mathbf{b}_1 + x_2 \mathbf{b}_2 + \cdots + x_N \mathbf{b}_N = 0.$$

Similarly, if the columns of a rectangular matrix of A are not independent, we have the nonzero solution vector $\mathbf{x}$ for

$$A\mathbf{x} \equiv \begin{pmatrix} a_{11} & a_{21} & \cdots & a_{M1} \\ a_{12} & a_{22} & \cdots & a_{M2} \\ \vdots & \vdots & \cdots & \vdots \\ \vdots & \vdots & \cdots & \vdots \\ a_{1N} & a_{2N} & \cdots & a_{MN} \end{pmatrix} \begin{pmatrix} x_1 \\ x_2 \\ \vdots \\ \vdots \\ x_M \end{pmatrix} = 0.$$

Thus we also have a nonzero solution vector for

$$A^{\mathrm{T}} A \mathbf{x} = 0.$$

The columns of the matrix $A^{\mathrm{T}} A$ are not independent when the columns of A are not independent. Thus, the matrix $A^{\mathrm{T}} A$ is not invertible when the columns of A are not independent.

9.5.3 Singular Value Decomposition

The eigenvalues of $A^{\mathrm{T}} A$ are nonnegative, so that

$$|A\mathbf{u}_i|^2 \equiv (A\mathbf{u}_i)^{\mathrm{T}} A\mathbf{u}_i = \mathbf{u}_i^{\mathrm{T}} A^{\mathrm{T}} A \mathbf{u}_i = \mathbf{u}_i^{\mathrm{T}} \sigma_i^2 \mathbf{u}_i = \sigma_i^2$$

where $\mathbf{u}_i$ denotes the ith unit orthogonal eigenvector and σ_i^2 denotes the ith eigenvalue of the matrix $A^{\mathrm{T}} A$. Let us express the vector $A\mathbf{u}_i$ as

$$A\mathbf{u}_i \equiv \begin{pmatrix} a_{11} & a_{21} & \cdots & a_{M1} \\ a_{12} & a_{22} & \cdots & a_{M2} \\ \vdots & \vdots & \cdots & \vdots \\ \vdots & \vdots & \cdots & \vdots \\ a_{1N} & a_{2N} & \cdots & a_{MN} \end{pmatrix} \begin{pmatrix} u_{i1} \\ u_{i2} \\ \vdots \\ u_{iM} \end{pmatrix} \equiv \sigma_i \mathbf{v}_i \equiv \sigma_i \begin{pmatrix} v_{i1} \\ v_{i2} \\ \vdots \\ \vdots \\ v_{iN} \end{pmatrix},$$

where $\mathbf{v}_i$ is the unit vector in the column space of A. The vectors $\mathbf{v}_i$ and $\mathbf{v}_j$ are orthogonal, since

$$\mathbf{v}_j^{\mathrm{T}} \mathbf{v}_i = \frac{1}{\sigma_j} (A\mathbf{u}_j)^{\mathrm{T}} \frac{1}{\sigma_i} A\mathbf{u}_i = \frac{1}{\sigma_j} \frac{1}{\sigma_i} \mathbf{u}_j^{\mathrm{T}} A^{\mathrm{T}} A \mathbf{u}_i = \frac{\Sigma_i}{\sigma_j} \mathbf{u}_j^{\mathrm{T}} \mathbf{u}_i = 0.$$

Suppose that the number of nonzero eigenvalues of the matrix A^TA is r. Taking the eigenvectors for the r non zero eigenvalues, we have the expression

$$A(\mathbf{u}_1 \quad \dots \quad \mathbf{u}_r) \equiv AU = (\mathbf{v}_1 \quad \dots \quad \mathbf{v}_r) \begin{pmatrix} \sigma_1 & \dots & 0 \\ \vdots & \ddots & \vdots \\ 0 & \dots & \sigma_r \end{pmatrix} \equiv V\Lambda,$$

where

$$\Lambda \equiv \begin{pmatrix} \sigma_1 & \dots & 0 \\ \vdots & \ddots & \vdots \\ 0 & \dots & \sigma_r \end{pmatrix}.$$

The rectangular matrix A is diagonalized using the orthogonal eigenvectors $\mathbf{u}$ and $\mathbf{v}$ as

$$AUU^T = A$$

$$= (\mathbf{v}_1 \quad \dots \quad \mathbf{v}_r) \begin{pmatrix} \sigma_1 & \dots & 0 \\ \vdots & \ddots & \vdots \\ 0 & \dots & \sigma_r \end{pmatrix} (\mathbf{u}_1 \quad \dots \quad \mathbf{u}_r)^T = V\Lambda U^T.$$

This diagonalization is called singular value decomposition of a matrix.
Thus we can obtain the relationship

$$A^TA = (V\Lambda U^T)^T(V\Lambda U^T) = U\Lambda V^TV\Lambda U^T = U\Sigma U^T$$

$$= (\mathbf{u}_1 \quad \dots \quad \mathbf{u}_r) \begin{pmatrix} \sigma_1^2 & \dots & 0 \\ \vdots & \ddots & \vdots \\ 0 & \dots & \sigma_r^2 \end{pmatrix} (\mathbf{u}_1 \quad \dots \quad \mathbf{u}_r)^T,$$

where

$$\Sigma \equiv \begin{pmatrix} \sigma_1^2 & \dots & 0 \\ \vdots & \ddots & \vdots \\ 0 & \dots & \sigma_r^2 \end{pmatrix}.$$

Similarly, we get

$$AA^T = V\Sigma V^T.$$

Thus, we can see the matrices A^TA and AA^T are orthogonally diagonalized, where U and V are orthogonal matrices whose columns are orthogonal eigenvectors of A^TA and AA^T, respectively, σ_i is the nonzero singular value of the matrix A, Σ is the diagonal eigenvalue matrix of A^TA and AA^T, and r is the number of nonzero eigenvalues for A^TA and AA^T. Table 9.5.1 illustrates an example of the singular value decomposition for a rectangular matrix.

Table 9.5.1 An example of singular decomposition of a rectangular matrix.

$$A = \begin{pmatrix} 4 & 11 & 14 \\ 8 & 7 & -2 \end{pmatrix} \qquad A^TA = \begin{pmatrix} 4 & 8 \\ 11 & 7 \\ 14 & -2 \end{pmatrix} \begin{pmatrix} 4 & 11 & 14 \\ 8 & 7 & -2 \end{pmatrix}$$

$$\begin{matrix} 7 & 7 \\ -1 & -1 \end{matrix} \qquad = \begin{pmatrix} 80 & 100 & 40 \\ 100 & 170 & 140 \\ 40 & 140 & 200 \end{pmatrix}$$

$$\begin{aligned} \sigma_1^2 &= 360 & \mathbf{u}_1 &= \left(\tfrac{1}{3}\ \tfrac{2}{3}\ \tfrac{2}{3}\right)^T \\ \sigma_2^2 &= 90 \\ \sigma_3^3 &= 0 & \mathbf{u}_2 &= \left(-\tfrac{2}{3}\ -\tfrac{1}{3}\ \tfrac{2}{3}\right)^T \end{aligned} \Bigg\} \ \mathbf{u}_1 \cdot \mathbf{u}_2 = 0$$

Not independent columns!! $A\mathbf{x} = 0$

No $(A^TA)^{-1}$ available $A^TA\mathbf{x} = 0$

3-dim row space   Orthonomal

$$U = \begin{pmatrix} \tfrac{1}{3} & -\tfrac{2}{3} \\ \tfrac{2}{3} & -\tfrac{1}{3} \\ \tfrac{2}{3} & \tfrac{2}{3} \end{pmatrix} \qquad \Lambda = \begin{pmatrix} 6\sqrt{10} & 0 \\ 0 & 3\sqrt{10} \end{pmatrix}$$

2-dimensional column space

$$V = \begin{pmatrix} 3/\sqrt{10} & 1/\sqrt{10} \\ 1/\sqrt{10} & -3/\sqrt{10} \end{pmatrix} \qquad A = V\Lambda U^T$$

Orthonomal

$$A = V\Lambda U^T = \begin{pmatrix} 3/\sqrt{10} & 1/\sqrt{10} \\ 1/\sqrt{10} & -3/\sqrt{10} \end{pmatrix} \begin{pmatrix} 6/\sqrt{10} & 0 \\ 0 & 3/\sqrt{10} \end{pmatrix} \begin{pmatrix} \tfrac{1}{3} & \tfrac{2}{3} & \tfrac{2}{3} \\ -\tfrac{2}{3} & -\tfrac{1}{3} & \tfrac{2}{3} \end{pmatrix}$$

$$= \begin{pmatrix} 18 & 3 \\ 6 & -9 \end{pmatrix} \begin{pmatrix} \tfrac{1}{3} & \tfrac{2}{3} & \tfrac{2}{3} \\ -\tfrac{2}{3} & -\tfrac{1}{3} & \tfrac{2}{3} \end{pmatrix} = \begin{pmatrix} 4 & 11 & 14 \\ 8 & 7 & -2 \end{pmatrix}$$

9.5.4 Pseudo-inverse Matrix

We can define the pseudo-inverse matrix using the singular value decomposition

$$A^+ \equiv (V\Lambda U^T)^+ \equiv U\Lambda^{-1}V^T.$$

This pseudo-inverse matrix satisfies the relationships

$$A^+A = U\Lambda^{-1}V^TV\Lambda U^T = UU^T$$

and

$$AA^+ = V\Lambda U^TU\Lambda^{-1}V^T = VV^T.$$

Here $A\mathbf{u}_i = \sigma_i\mathbf{v}_i$. Thus $\mathbf{u}_i$ is a vector in the row space and $\mathbf{v}_i$ is a vector in the column space. The matrix AA^T is called the projection matrix since

$$AA^T\mathbf{x} \equiv (\mathbf{a}_1 \quad \ldots \quad \mathbf{a}_M)\begin{pmatrix} \mathbf{a}_1^T \\ \vdots \\ \mathbf{a}_M^T \end{pmatrix}\mathbf{x} = \begin{pmatrix} \mathbf{a}_1\mathbf{a}_1^T\mathbf{x} \\ \vdots \\ \mathbf{a}_M\mathbf{a}_M^T\mathbf{x} \end{pmatrix}.$$

The vector $\mathbf{x}$ is projected onto the space of $(\mathbf{a}_1 \ldots \mathbf{a}_M)$. Thus, the matrix UU^T is the projection matrix onto the row space of A and VV^T is the matrix for the projection onto the column space of A.

9.6 Pseudo-inverse Matrix and Linear Equations

9.6.1 Minimum-norm Solution and a Null Space

The pseudo-inverse matrix can be applied to solve a set of simultaneous equations. The linear equation $A\mathbf{x} = \mathbf{b}$ has a unique solution $\mathbf{x}$ when the matrix A has its inverse A^{-1}. Even if the matrix does not have its inverse A^{-1}, we can take the pseudo-inverse A^+ to get the appropriate solution.

Let us take the pseudo-inverse of A. We can write the solution as

$$\mathbf{x}^+ \equiv A^+A\mathbf{x} = UU^T\mathbf{x},$$

where $A = V\Lambda U^T$ and $A^+ \equiv (V\Lambda U^T)^{-1}$. Here the solution $\mathbf{x}^+$ is the projected vector onto the row space.

If the matrix A has a null space, that is, there is a vector $\mathbf{x}_0$ that satisfies the homogeneous equation $A\mathbf{x}_0 = \mathbf{0}$, then the solution for $A\mathbf{x} = \mathbf{b}$ is not unique. That is, the matrix A does not have its inverse A^{-1}. When the column vectors of the matrix A are not independent, the null space exists. This is because we can find the vector $\mathbf{x}_0$ which satisfies the homogeneous equation as

$$A\mathbf{x}_0 \equiv x_{01}\mathbf{a}_1 + \cdots + x_{0M}\mathbf{a}_M = \mathbf{0}.$$

The null space is orthogonal to the row space which is spanned by the row vectors, since we can rewrite the homogeneous equation as

$$A\mathbf{x}_0 = \begin{pmatrix} \mathbf{b}_1^T \\ \vdots \\ \mathbf{b}_M^T \end{pmatrix}\mathbf{x}_0 = \mathbf{0}$$

where $\mathbf{b}_i^T$ denotes the ith row vector of A.

The solution vector $\mathbf{x}$ for $A\mathbf{x} = \mathbf{b}$, where the matrix A has a null space, can be decomposed into the projection vectors onto the row and null spaces, since these two spaces are orthogonal. Then we can define the minimum-norm solution as the row space projection vector. Suppose that we have a set of simultaneous equations (Lay, 1994),

$$x_1 + 2x_2 + 3x_3 = 1,$$
$$2x_1 + 4x_2 + 8x_3 = 6,$$
$$3x_1 + 6x_2 + 11x_3 = 7.$$

A particular solution set is given by

$$\mathbf{x}_p \equiv \begin{pmatrix} -5 - 2x_2 \\ x_2 \\ 2 \end{pmatrix},$$

where x_2 is a free parameter. The homogeneous solutions for

$$x_1 + 2x_2 + 3x_3 = 0,$$
$$2x_1 + 4x_2 + 8x_3 = 0,$$
$$3x_1 + 6x_2 + 11x_3 = 0,$$

are given by

$$\mathbf{x}_0 \equiv x_2 \begin{pmatrix} -2 \\ 1 \\ 0 \end{pmatrix}.$$

Thus the complete solution $\mathbf{x}$ becomes

$$\mathbf{x} = \mathbf{x}_p + \mathbf{x}_0 = \begin{pmatrix} -5 \\ 0 \\ 2 \end{pmatrix} + x_2 \begin{pmatrix} -2 \\ 1 \\ 0 \end{pmatrix} \begin{matrix} \equiv \mathbf{X}_{PM} + \mathbf{X}_{OM} \\ \equiv \mathbf{X}_{PM} + \mathbf{X}_{null} \end{matrix}.$$

This solution vector can be illustrated as shown in Fig. 9.6.1. The minimum-norm solution $\mathbf{X}_{PM}$ is obtained by the projection of the complete solution onto the row space. After subtraction of the null-space component of the complete solution $\mathbf{X}_{null}$, we can get the minimum-norm solution. The orthogonal projection of the complete solution onto a null-space solution vector is

$$\frac{1}{5} \begin{pmatrix} -7 \\ 1 \\ 2 \end{pmatrix}^{\mathrm{T}} \begin{pmatrix} -2 \\ 1 \\ 0 \end{pmatrix} = 3;$$

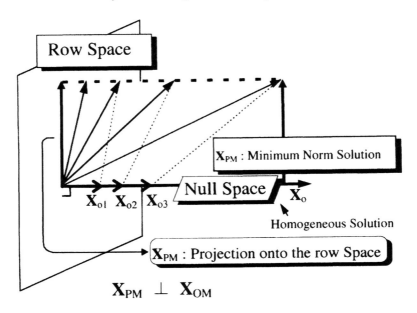

Figure 9.6.1 Geometric interpretation of minimum-norm solution and complete solution set.

then we get the null-space component for the complete solution as

$$\mathbf{x}_{null} = 3 \begin{pmatrix} -2 \\ 1 \\ 0 \end{pmatrix}$$

when taking $x_2 = 1$. Thus the minimum-norm solution is given by

$$\mathbf{X} - \mathbf{X}_{null} \equiv \mathbf{x}_{PM} = \begin{pmatrix} -7 \\ 1 \\ 2 \end{pmatrix} - 3 \begin{pmatrix} -2 \\ 1 \\ 0 \end{pmatrix} = \begin{pmatrix} -1 \\ -2 \\ 2 \end{pmatrix}.$$

This is the projection of the complete solution onto the row space.

Thus, if we take the pseudo-inverse A^+ of the matrix A, we can get the solution as

$$\mathbf{x}^+ \equiv A^+\mathbf{b} = A^+A\mathbf{x} = UU^{\mathrm{T}}\mathbf{x},$$

where $A = V\Lambda U^{\mathrm{T}}$. The solution $\mathbf{x}^+$ is the projection of the solution $\mathbf{x}$ onto the row space. This solution also satisfies the equation

$$A\mathbf{x}^+ = AA^+\mathbf{b} = AA^+A\mathbf{x} = V\Lambda U^{\mathrm{T}}UU^{\mathrm{T}}\mathbf{x} = A\mathbf{x} = \mathbf{b}.$$

Therefore, the pseudo-inverse gives the minimum-norm solution when the matrix has a null space, or equivalently, when the linear equation does not have a unique solution.

9.6.2 The Least Square Error Solution

Again suppose that we have a linear equation

$$Ax = b,$$

where $A = V \Lambda U^\mathrm{T}$. Let us take a solution as the least square error solution

$$\hat{x}^+ \equiv A^+ \hat{b},$$

where $\hat{b}$ denotes the projection vector of b onto the column space.

The solution $\hat{x}^+$ can be obtained by directly applying the pseudo-inverse matrix with the vector b instead of $\hat{b}$. That is,

$$\hat{x}^+ \equiv A^+ \hat{b} = A^+ A A^+ b = A^+ V V^\mathrm{T} b = U \Lambda^{-1} V^\mathrm{T} V V^\mathrm{T} b$$
$$= U \Lambda^{-1} V^\mathrm{T} b = A^+ b \equiv x^+.$$

The solution $\hat{x}^+$ satisfies the equation

$$A\hat{x}^+ = Ax^+ = A A^+ b = \hat{b},$$

where $A A^+ b$ shows the projection vector of b onto the column space.

Thus, when the vector b is not located in the column space of A, the pseudo-inverse gives the solution for the projection vector of b onto the column space. That is, we can obtain the solution $x^+ = A^+ b$ for the linear equation

$$Ax^+ = \hat{b}$$

where $\hat{b}$ denotes the projection vector of b onto the column space.

9.6.3 Minimum-norm Least Square Error Solution

We have already tried to get the least square error solution for the equation $Ax = b$ by solving the normal equation

$$A^\mathrm{T} A \hat{x} = A^\mathrm{T} b,$$

where we assumed that the matrix $A^\mathrm{T} A$ is invertible. However, the matrix $A^\mathrm{T} A$ is not always invertible. When the columns of the matrix A are not independent, the matrix $A^\mathrm{T} A$ is not invertible. Therefore, we cannot obtain a unique solution by solving the normal equation.

Let us rewrite the equation which the least square error solution satisfies as

$$A\hat{x} = \hat{b},$$

where $\hat{\mathbf{b}}$ is the projection of the vector $\mathbf{b}$ onto the column space. The normal equation does not have a unique solution when the equation above does not provide a unique solution.

The solution obtained by taking the pseudo-inverse,

$$\hat{\mathbf{x}}^+ \equiv A^+\hat{\mathbf{b}},$$

gives us the minimum-norm solution for such a case. And the minimum-norm solution above for the equation $A\hat{\mathbf{x}} = \hat{\mathbf{b}}$ gives us the minimum-norm and least square error solution for the equation $A\mathbf{x} = \mathbf{b}$.

The solution $\hat{\mathbf{x}}^+ \equiv A^+\hat{\mathbf{b}}$ can be rewritten as

$$\hat{\mathbf{x}}^+ \equiv A^+\hat{\mathbf{b}} = A^+AA^+\mathbf{b} = A^+VV^+\mathbf{b} = U\Lambda^{-1}V^\mathrm{T}VV^\mathrm{T}\mathbf{b}$$
$$= U\Lambda^{-1}V^\mathrm{T}\mathbf{b} \equiv A^+\mathbf{b}.$$

Thus, the solution $\hat{\mathbf{x}}^+ = A^+\hat{\mathbf{b}}$ can be obtained by applying the pseudo-inverse to vector $\mathbf{b}$. The solution $\hat{\mathbf{x}}^+ \equiv A^+\hat{\mathbf{b}}$ satisfies

$$\hat{\mathbf{x}}^+ \equiv A^+\hat{\mathbf{b}} = A^+A\hat{\mathbf{x}} = UU^\mathrm{T}\hat{\mathbf{x}}$$

and

$$A\hat{\mathbf{x}}^+ = AA^+\hat{\mathbf{b}} = \hat{\mathbf{b}}.$$

Therefore, the solution $\hat{\mathbf{x}}^+ \equiv A^+\hat{\mathbf{b}}$ is the minimum-norm least square solution for the equation $A\mathbf{x} = \mathbf{b}$.

9.7   Summary

We have described the linear equation models for inverse filters, signal analysis, linear and orthogonal regressions, and principal component analysis. Signal processing using a finite number of data records can be formulated by linear equations. And the least square error criterion is quite an important basis for solving the linear equations which are employed in acoustic signal processing. The pseudo-inverse matrix using the singular value decomposition is important, particularly for finding the least square error solution. The elements of matrix theory are not described in this chapter. Readers who are interested in acoustic signal processing are encouraged to study the fundamentals of the mathematical theories of the linear equations. A standard textbook such as *Linear Algebra and its Applications*, by David C. Lay (Addison-Wesley, 1994) is recommended.

References

Allen, J. and Berkley, D. (1979). *J. Acoust. Soc. Am.*, **65**, 943–950.

Lamb, Sir. H. (1945), *Hydrodynamics*, Dover, New York.

Lay, D. (1994), *Linear Algebra and Its Applications*, Addison-Wesley, Reading, Mass.

Lyon, R. (1969), *J. Acoust Soc. Am.*, **45**, 545–565.

Lyon, R. (1983), *J. Acoust. Soc. Am.*, **73**, 1223–1228.

Lyon, R. (1984), *J. Acoust. Soc. Am.*, **76**, 1433–1437.

Lyon, R. (1987), *Machinery Noise and Diagnostics*, Butterworths, Boston.

Mathews, J. and Waker, R. (1973), *Mathematical Methods of Physics*, W.A. Benjamin, Menlo Park, California.

Morse, P. and Bolt, R. (1994), *Rev. Mod. Phys.*, **16**, 69–150.

Morse, P.M. and Ingard, K.U. (1968), *Theoretical Acoustics*, Princeton University Press, Princeton, N.J.

Rayleigh, J.W.S. (1945), *Theory of Sound*, Dover, New York.

Roach, G. (1970), *Green's Functions*, Van Nostrand Reinhold, New York.

Rossing, T. and Fletcher, N. (1995), *Principles of Vibration and Sound*, Springer-Verlag, New York.

Schroeder, R. (1989), *J. Audio. Eng. Soc.*, **37**, 795–808.

Terada, T. *et al*. (1994), *Proc. IEEE-Sp Int. Symp. TF/TS Analysis*, 429–432.

Tohyama, M., Suzuki, H., and Ando, Y. (1995), *The Nature and Technology of Acoustic Space*, Academic Press, London.

Waterhouse, R. (1963), *J. Acoust. Soc. Am.*, **35**, 1141–1151.

Index

Lightning Source UK Ltd.
Milton Keynes UK
UKOW03n1523170913

217329UK00010B/302/A

9 780126 926606